BRITISH BEE BOOKS

a bibliography 1500–1976

Mel is hony and is hote in ye fyrst degre/ & drye in the secōde/ hony is made by artyfyce/ and craft of bees. The whyche bees draweth the thynnest parte of the floures/ and partely of the thickest & moost grosse/ and therof maketh hony and ware/ and also they make a substaunce that is called the hony combe.
The tame hony is that that is made in the hous or hyues that labourers ordeyneth for the sayd bees to lodge and worke in.
Hony is whyte in colde places/ & browne in warme places. And hony ought to be put in medicyns and may be kept. C. yeres
There is an other that is called wylde hony and is founde in woodes/ and is not so good as the other and is more bytter/ and therfore it is put in medycyns to unstoppe the cōduytes of uryne/ & is better therfore than the other. Also there is a hony called castanea bycause it is made of chestayne floures that the bees sucketh/ and is bytter and dyurytyke as the other Hony hath vertue to consume/ to clēse/ and kepe thynges fro rotynge. And is put in medycyns to delay and hyde the bytternesse of medycyns/ and that the medycyns gō to the botom of membres bycause of the swetnesse therof/ that is medled with them. Powdre of medycyns is medled with hony to kepe the vertue therof.

¶ For the stomake. A

¶ Agaynst colde humours in the stomake Take a drynke called mulsa that is of hony and warme water.

¶ For the herte. B

¶ For weykenesse of the herte and swownynge. Take ye brothe made of hony with some water that conforteth.

¶ To clense the stomake. C

¶ To clense dyrte or fylthe in the stomake and to unbynde the bely. Take nytre with hony and it clenseth.

¶ For euyll colour in the face. D

¶ For euyll colour in the face called pannus that cometh to womē after theyr chyldynge. Confyct two or thre dragme of nytre with hony/ and lete it lye the space of thre dayes/ and therwith anoynte the face And for the same hony with bulles galle is good.

¶ For to unbynde the bely. E

¶ In a feuer or other dysease whā the bely is bounde/ a supposytory is made in this maner. Hony is sodē in a vessel tyll it ware blacke/ and powdre of salte is put therto/ and whan they be medled the hony is poured on a stone or an other euen thynge/ and than be anoynted as a supposytory ought to be/ and put it in the foundement/ and it wyll cause you to shyte a pace.

The section on bees from *The grete herball* published in 1526 (entry 2), with the first illustration of bees and hives to be printed in England. Reproduced by permission from the copy of the book in the British Library.

BRITISH BEE BOOKS

A BIBLIOGRAPHY 1500–1976

by

International Bee Research Association

Principal collaborators

Joan P. Harding, F.S.A.
David A. Smith, M.A.
Eva Crane, M.Sc., Ph.D.
Rosamund M. Duruz, M.A.

International Bee Research Association
London
1979

British Library Cataloguing in Publication Data

International Bee Research Association
British bee books.
1. Bee culture—Bibliography
I. Title II. Harding, Joan P.
016.638'1 25256

ISBN 0-86098-077-4

Style of reference recommended for entries in this ***Bibliography***:
IBRA 281, IBRA *ms* 7

Printed by Unwin Brothers Limited
The Gresham Press, Old Woking, Surrey

Contents

List of Illustrations

In the next year, 1887, Samuel Simmins' book *A modern bee farm* was published in both a popular edition and a special edition with sepia photographs (see note to 387). From this time on, photographs were increasingly used to illustrate bee books, and many recent books contain superb photographs in colour. The cost of reproducing a selection of these would be prohibitive, and we therefore leave the series of illustrations before the art of line-engraving declined, adding one earlier example for entertainment rather than instruction.

Introduction

by Eva Crane
Director, International Bee Research Association

When book publishing began

British bee books are a national, and indeed a world, heritage with a very long history. When the first book was printed in England in 1477, there had been no advancement in beekeeping since the time of Ancient Rome, more than a thousand years earlier. Indeed the first printed bee books lacked many of the refinements in method and materials used by some of the leading Roman beekeepers.

Beekeeping was almost certainly widespread in Britain and Ireland in the 1400s. Wicker basket hives were used, except in the south and east where baskets of coiled straw or grass had been introduced by the Saxons who had settled there. Both types of hive are known as skeps. They were stood off the ground on stools, or on a bench or some other support. Bee boles (recesses for skeps) that were built into a wall near Corfe Castle in Dorset in the fourteenth-century still survive.

At the end of each summer, the honey was taken from the lightest hives, which would have least chance of survival until next spring, and also from the heaviest, which would yield most honey. The bees in these hives were killed, for instance by setting each hive in turn over a hole dug in the ground where sulphur had been set alight.

The important operations in the beekeeping year were the honey harvesting, feeding the stocks to be overwintered if this was done, setting the hives outside in the spring if they had been kept in an outhouse or cellar while the weather was too cold for the bees to fly, and finally swarm management in early summer. The swarms were needed to repopulate the hives whose bees had been killed off. Early swarms were very valuable, because they still had time to build up their population and to store honey; late swarms were worthless because they could not do this. Hence the rhyme:

A swarm of bees in May
Is worth a load of hay.
A swarm of bees in June
Is worth a silver spoon.
A swarm in July
Is not worth a fly.

Much attention was paid to dates in the beekeeping calendar (entry 3), as in other agricultural pursuits.

One type of beekeeping, which seems to have been indigenous to the forests of northern Europe, owed little to the Roman and earlier Mediterranean cultures. From very early times honey had been harvested from swarms of bees that built their nests in hollow trees. Such bee trees were highly valued, and rights of ownership grew up, the owner tending the colonies of bees in his own trees. Later, the trunk containing a bees' nest would be cut down, and used as a separate hive that could be moved near the owner's dwelling for protection. There is only the scantiest evidence that such hives were used in Britain or Ireland, but the rights of interested parties to the bees, honey and wax in *wild* nests were set out in great detail in the laws of Ireland, Wales and the royal forests of England (*ms* 1, *ms* 2, *ms* 9).

Five hundred years of printed books

The earliest books preserved the beliefs and practices of the past. The first true English beekeeping book 'written out of experience' is Edmund Southerne's *Treatise* (12) published in 1593. It is a modest enough beginning, to be followed only 16 years later by one of the greatest of all British beekeeping books, *The feminine monarchie* (18) by the Rev. Charles Butler (1609), which also marks the beginning of bee science. Butler wrote a detailed and scholarly description of skep beekeeping; he rejected the absurdities of the Roman writers, and he made original observations, for instance that the drones were males. He stated that the queen was a female; in Spain, Luis Méndez de Torres had in fact established this, and in 1586 reported in his book that the leader bee in the hive laid the eggs from which all workers, drones, and future queens developed. No evidence has come to light that Butler knew of this book, and he thought the workers laid the eggs.

The 1600s witnessed an unparalleled flowering of scientific enquiry into the life history of bees and into better ways of managing them. Men of science in Britain and continental Europe worked on bees, and bee books were written in England by Richard Remnant, Samuel Purchas, Samuel Hartlib, John Worlidge, John Gedde, and the reactionary Moses Rusden. This century was one of the most interesting periods in the history of man's relationships with bees. The position was consolidated in the 1700s, with translations of works by Réaumur and Swammerdam, and books describing practical beekeeping experience that presented new ideas, such as those by Joseph Warder, John Thorley, Stephen White, Thomas Wildman and James Bonner.

The 1800s can be neatly divided into two half-centuries. In the first, Thomas Nutt and William Cotton were among those who led a popular movement against the killing of bees when harvesting their honey. They devised methods of using a separate section of the hives as a specific honey chamber, from which all the bees could be 'driven' when honey was harvested, so that they rejoined the brood chamber which was left intact. Bar-hives were developed in which the bees built their combs in an orderly fashion, hanging from top-bars laid across the top of the hive. During this time also there were many exhortations to cottagers and others to keep bees, and help and advice were provided for them.

Then in 1851 came the culmination of thought and effort, in many countries, directed towards more complete management of the bees. In that year the Rev. L. L. Langstroth appreciated the significance of the 'bee space' and the fact that bees will not build their combs across a gap of this size—about 6 mm. Working in the U.S.A., he devised a way of providing hanging frames that the bees would build their combs in, and produced the first movable-comb hive that was practical enough to gain world-wide acceptance.

The man who introduced this movable-frame hive into Britain was T. W. Woodbury of Exeter, Devon. A few years ago Graham Burtt of Gloucester came across the copy of Langstroth's book *The hive and the honey bee* that the author had presented to Woodbury, and obtained it for the I.B.R.A. Library. This book brought 'rational' beekeeping to Britain, and it was followed by many others giving instructions for the new method, and a few arguing against it. There was much more to learn about beekeeping now: beekeeping associations and journals proliferated, to exchange experience and to give advice and instruction, and they themselves published books.

By the turn of the century some quite large beekeeping enterprises had grown up, and these were described in books by their proprietors. All manner of people became interested in beekeeping, and more than a few produced books: hive manufacturers, schoolmasters, doctors, clergymen, and government experts. It is worth noting how many books in this *Bibliography* were published by their authors, without the discipline that must be imposed by a publishing house. Many beekeepers are adventurous, and individualists. Since their bees, being very adaptable, have a way of surviving, each can think that his own method is the best, and to some extent this is true, because the method must suit the person as well as the bees.

After 1930 the books come thick and fast, and now there are a good many new bee books each year. The exact date of publication is of less importance than the name of the author. The books published in the last fifty years are therefore listed by author instead of by date. Specialization has continued, and a Subject Index is included on page 267 to help the reader to find what he wants. But there has been no reduction in the number of general books on bees and beekeeping, and all ages and all educational backgrounds are catered for.

The proliferation of books for the young has been a feature of recent years, and a *Bibliography of children's books on bees and beekeeping* was published in 1973 (665). Since the end of the Second World War there has been a great expansion of scientific work on bees, which are easily kept for experimental use, and which are of great interest to the student of animal behaviour because of their advanced social structure. Interest in and knowledge about the various hive products, and the biochemistry and physiology of their production by the bees, have prospered. Beekeepers and bee scientists are kept informed about all these current developments through the International Bee Research Association and its journals. Activities connected with historical material are much more difficult to finance, and all the early beekeeping books in the I.B.R.A. Library were gifts or bequests. If readers see the possibility of adding to our Library any of the books or editions not marked B in the entries, we should warmly welcome the opportunity of acquiring them.

The present and future

The primary medium for information on current research and development is the journal *Apicultural Abstracts*, and the *Index to Apicultural Abstracts* 1950–1972 (604) provides a computer-generated guide to this modern knowledge and expertise, as well as to writings on beekeeping practice and its technology. The *Index* shows that there are still gaps among the books that beekeepers and bee scientists need today. There is no adequate book on beeswax; much more is known about beeswax now than in 1908 when the last book was published in Britain about it, and the book by H. H. Root, published in the U.S.A. in 1951, is far from adequate. A successor to Dr. C. R. Ribbands' *Social life and behaviour of honeybees* (752) is needed, as a guide to what is now known about the subject. But much time and effort are expended in writing a readable synthesis of recent research work, and since this is likely to be soon outdated, its preparation may seem an unrewarding task. At the other end of the scale, facsimile reprints of early books, out of copyright, constitute an almost effortless form of book publishing that could well be extended.

Consideration of the influence of any group of specialist books must take into account the question of language. Bees are kept in virtually every country where books are published, and publications on bees and beekeeping have been printed in at least 58 languages. English has always been a leading language for bee books, and in the past 20–30 years its importance has increased even further. No figures are available for books as such, but of the world's current scientific publications on bees and beekeeping 70% are likely to be in English, and not more than 5% in any other single language except Russian.

In publishing a bibliography of past bee books, it is proper to consider books that will be published in the future, at any rate in the next few decades. Because of the importance of the English language, a definitive book in English is needed on every specialized subject. Many such books are published in Britain and in the U.S.A., including books on bees and beekeeping. So far there has been rather little joint production of books between the two countries, and such a venture could usefully be explored. Inter-continental collaboration is easier nowadays than it has ever been, and it is also more necessary: bees (and their parasites that can be sent with them by airmail post) have become a global concern.

A century ago beekeepers were living in the exciting period when new frame-hive beekeeping was being promulgated in their own countries, and their books reflect this situation. Readers of the present book were born too late to share that expansion of ideas and interest at first hand, but there is a current development that in some ways parallels it, and that will influence future British bee books. A hundred or more developing countries in the tropics are actively concerned to increase and improve their beekeeping, and the great majority of these countries use books in English. Moreover many of them have a local race or species of honeybee that evolved in the tropics, about which comparatively little is known, and still less is published in books. The African and Asiatic honeybees are scantily documented in comparison with the European bees that formed the basis of last century's beekeeping advances. In the next half-century these tropical bees, and their management, are also likely to be important features of books published in Britain.

In the next few decades, therefore, we should be able to look forward to new British bee books at all levels, from children's illustrated texts to scientific and scholarly handbooks, and from manuals on beekeeping in different regions of Britain to surveys relating to the temperate zones, subtropics and tropics, and to beekeeping as a world-wide interest. Beekeepers being great individualists, there will surely be some unexpected and innovative books as well.

Preface

by David A. Smith
Secretary, International Bee Research Association

Scope

It has been the aim of the *Bibliography* to list every book with more than 20 pages on bees, published in Great Britain and Ireland from earliest times to 1976. It should thus provide a useful guide to librarians, to the many collectors of bee books, and to specialist booksellers, as well as to interested beekeepers. It also presents a picture of the historical development of beekeeping and bee research, not only in Britain but also throughout the world. To complete this picture, particularly for the earlier years, other books with references to bees are included—herbals, books on agriculture, books on insects—and we have upon occasion broken the rule and included works of less than 20 pages, or more general works with less than 20 pages dealing with bees. Certain manuscripts before 1500 are listed although they contain quite short passages on bees, because they show the state of knowledge at the time. We have also included in the main list certain later unpublished manuscripts which are known to us (e.g. 30, 44, 499a).

The entry for each book published up to 1926 indicates libraries that have a copy. For the I.B.R.A. Library this is continued up to 1976, so the *Bibliography* also serves as an annotated catalogue of British bee books in this Library.

Because the subject of the *Bibliography* is British bee books, not all English editions of classical authors are listed, although even up to the seventeenth century they were regarded as important authorities. The best known translations of Virgil's *Georgics* are however included, together with some other early translations from classical authors. Articles in periodicals are listed only up to 1800.

A few works of fiction are included, which can reasonably be described as bee books (if they are about bees, e.g. 689, or if bees play a major part in the story, 522, 717). The fact that A. A. Milne's famous story *In which we are introduced to Winnie-the-Pooh and some bees, and the stories begin* appeared in 1970 as a book on its own, justifies its inclusion (720, 721). Mrs. Gatty's *Parables from Nature* (279) is perhaps more difficult to justify.

Other stories which mention bees, such as Rudyard Kipling's *Dymchurch flit* from *Puck of Pook's Hill*, *The mother hive* from *Actions and reactions*, and Roald Dahl's *Royal jelly* from *Kiss-kiss*, are not listed because they are not books in their own right.

It is helpful to prospective purchasers of books to know if titles that have bee connotations are really about bees or not, and all such titles are therefore included in the *Bibliography*. Satires about bees, for instance *The Fable of the bees* (76) and Gay's *Fables* (91a), provide evidence of the knowledge of bees at the time of their publication. It has been difficult to draw the line between satires and other books whose titles suggest that they are about bees although the only link is the industry or some other attribute of bees. These are in general listed separately under 'metaphorical titles' on page 235.

No attempt has been made to include the many references to bees in English literature. Thus there is no entry for the famous speech, in Shakespeare's *Henry V* Act 1 Scene 2, beginning 'So work the honey-bees . . .' (1599), nor to the many passages mentioning bees in John Clare's *The shepherds calendar* (1827). Another absentee is the fictional book *Practical handbook of bee culture, with some observations upon the segregation of the queen* by Sherlock Holmes, described by him in *His last bow* as the *magnum opus* of his latter years.

We are conscious that the choice of entries shows many inconsistencies, partly because we have tended to include borderline books when we had information about them. At one time it was intended that the *Bibliography* should include all books in English, irrespective of the country of publication. In fact the list for Canada and the U.S.A. was prepared more quickly, and Dr. T. S. K. and Mrs. M. P. Johansson were able to publish it in 1972 (page 22). Preliminary lists for Australia, India and New Zealand are now in preparation by authors in these countries.

History and Acknowledgements

This work has had a long period of gestation. In the early 1950s Rosamund Duruz, who was then the Keeper of the Beekeeping Museum founded by the Bee Research Association, began to collect details of early British books on bees. The original intention was to cover only the period up to 1700, but the list was later extended to 1800. It was an important feature of the original plan that details should be given of the major libraries in Britain and Ireland which hold copies of the books, together with holdings of three major libraries in the U.S.A. Joan Harding, who had already been helping in the preparation of the list, took over the responsibility for it when Rosamund Duruz moved to Australia in 1954. By this time the list had been extended again, and there were thoughts of taking it as far as 1950.

This early progress was recorded in the B.R.A. Annual Reports for 1953 and 1954. Thereafter Joan Harding worked almost unaided on the *Bibliography*; she spent much time in the British Museum Library and others, examining books and checking references, and she produced a first typescript which was retyped on multiple reference slips by Phyllis Bingham, a member of B.R.A. staff. A great deal of checking and completing of entries remained to be done, but the work existed in embryo. The late Dr. H. M. Fraser's *History of beekeeping in Britain* published in 1958 contained the announcement: 'Shortly to be published by the Bee Research Association as a companion volume to this book, *British bee books: a bibliography 1500–1956* compiled by Joan Harding.' Dr. Fraser himself read through the slips, and some of his comments are included in the entries. Unfortunately the necessary work of checking and editing could not be completed in the years that followed, because of the continuous shortage of funds under which the Association operated. In 1974, the history of the first 25 years of B.R.A. stated (page 89) that 'the delay in completing this book . . . was in a sense a necessary sacrifice to the Association's struggle for survival.'

Meanwhile the index slips were in frequent use in the B.R.A. Library, and Eva Crane was able to make many further corrections and additions. The major participation of Dr. Crane is almost a *sine qua non* of any of the works produced by I.B.R.A., and the *Bibliography* is certainly no exception. It was she who inveigled me in 1964 into helping with the work. This involved examining many catalogues and books in libraries (especially the British Library) and at various booksellers. No one person has been able to examine every book described, and a few have not been seen at all; with many books it has not been possible to examine all editions. Some of the entries remain as they were in the original typescript of the 1950s, and in others both bibliographical details and notes have been entirely rewritten.

The difficulty of the International Bee Research Association (the name was changed in 1976) has been to find the time and financial resources to complete the preparation of the typescript to its final form, and to finance publication. A grant awarded in 1977 by the Development Commissioners towards the cost of printing has provided the necessary impetus to complete publication.

In recent years I have been chiefly responsible for the entries up to 1926, and Eva Crane has compiled the *Author bibliography for the last fifty years*, taking the entries up to 1976. However, the various compilers' responsibilities have overlapped considerably, and members of I.B.R.A. staff have also played an important part, in particular Audrey Pitman, Sue Morris and Chris Edwards.

It would not have been possible to produce this work without the assistance of many individuals and institutions, and we cannot mention all by name. Information given to us by the staff of many libraries has been integrated into the *Bibliography*, and particularly of those libraries whose holdings are listed, among whom special mention should be made of the staffs of the British Library and the Ministry of Agriculture Library, Catherine Dixon of the Moir Library in Edinburgh, and T. E. Cawley of Rothamsted Experimental Station. Other librarians who have given much help include A. E. Barker (S.P.C.K.), Gavin Bridson (Linnean Society), Kathleen Gilbertson (Steenbock Memorial Library, Wisconsin), William Linnard (Welsh Folk Museum), Sir Robin Mackworth-Young (the Queen's Librarian at Windsor Castle), Elizabeth McNamara (Trinity College Library, Dublin), H. T. Murphy (Albert R. Mann Library, Cornell University).

Many booksellers have allowed me to make notes of books on their shelves or in their catalogues, including Bee Books New and Old, Kew Books, Northern Bee Books, Michael Phelps, Bernard Quaritch Ltd, Wheldon & Wesley Ltd, Jeanette White. Assistance on various points has been given by publishers, especially Bell & Hyman Ltd., the Folio Society, E. H. Taylor Ltd., Theatrum Orbis Terrarum, and by the National Museum of Wales. Bishop Mark Hodson and T. E. Lewis are among the bibliophiles who let me consult their private collections of bee books.

It is also appropriate to record here our gratitude to the authors of the books listed on page 22, from whose works we have drawn freely (particularly the late Colonel H. J. O. Walker and Dr. H. M. Fraser), and our appreciation of the catalogues of the Rothamsted and Moir Libraries.

For many years the International Bee Research Association has received grant-aid from the Development Commissioners, and they have provided an additional grant to enable us to publish this book. Publication has also been supported by a number of Beekeepers' Associations in Britain: the Welsh Association, the County Associations of Hampshire, Hertfordshire, Oxfordshire, Yorkshire and Yorkshire (South), and the District Associations of Aberdeen, Chalfont, Enfield, Inner London, Manchester, N.W. Middlesex, Twickenham and Thames Valley, and also D. J. Deasy, President of the Irish Association.

Guide for users of the Bibliography

Sequence of entries

In all sections except the last (1927–1976), entries are in order of the year in which the first edition of the book was published. In a few instances where the first edition contains no bee material, or is unknown to us, a book may be listed under a later edition (e.g. entries 7, 70, 190). The only other exception is that manuscripts first published several centuries after they were written are entered under the year of the original manuscript (e.g. 28, 34). Books published in the same year are usually entered alphabetically by author. A few exceptions have been made, for instance to list consecutively two volumes of the same series or two books by the same author.

In the final section (1927–1976) entries are arranged alphabetically by author, and books by the same author are entered in order of the year of publication.

Content of entries

Component 1: author, entry number, date. The author's name and style are usually copied from the title page.

Component 2: title (short title in italics)/town of publication (county in brackets if useful), imprint or publisher, date./pagination, details of illustrations, bibliography./libraries holding the book (see *Key* below), those in the U.S.A. being entered last. In the pagination, if the book has only sections on bees, the page numbers are indicated, either as Bees pp. 52–78 or Of the hony-bee . . . pp. 114–124; here 'Of the hony-bee' is the title of the section occupying pp. 114–124. The full title is normally given, except for some early books with only a section on bees. Arabic numerals are used for the main pagination, Roman for preliminary pages; the numbering of plates is copied from the book where possible. The page size (height) is given if it is not between 16 and 24 cm.

Component 3: Information about editions after the first (full details only for books entirely about bees).

Component 4: Entry numbers in other bibliographies (see *Key* below), e.g. Hodgson 267. A Hunt or Cole entry number implies that the book is in the Hunt or Cole Library, respectively (see *Key*).

Component 5: Notes of interest, normally relating to all editions; these are fuller for early books.

The final section (1927–1976) normally gives components 1–4 only, and Component 4 is restricted to Hodgson and Johansson. Fiction, verse and children's books are, however, designated. Only one library letter is given (B or an alternative), and we have not established a library holding for every edition of every book.

Square brackets are used in components 1–3 to indicate information not given in the book itself.

Shortcomings

During the twenty-five years of preparation, entries have been compiled by many people, and inconsistencies certainly remain, for instance in the method of calculating pagination, and in information about illustrations, specific details of which are given only for editions inspected recently; only the general word 'illus.' is used for others, as for inspected books with line drawings in the text only. The use of square brackets is occasionally copied from a catalogue in which their implication is not clear to us, and there are other details from different library catalogues which cannot be presented in a consistent style.

We have adopted the policy of including information where we have it, even if similar information cannot be guaranteed in all appropriate entries. For instance it is certain that libraries in our list hold other books in the *Bibliography* than those marked with the appropriate letter. Information included for one book or edition (e.g. dedicated to . . .) may not be given for another book to which it also applies. The *Bibliography* is not a catalogue of a single library whose books could, or can, be seen and compared together and referred to again and again as necessary, but a distillation from many sources, and it is offered as it now stands, 'with all faults'.

Key to abbreviations and symbols

Library addresses are given on page 21.

A

A copy in British Library, London
anr. another

B

B copy in International Bee Research Association Library
(B) photocopy in I.B.R.A. Library
B.B.J. *British Bee Journal*
B.B.K.A. British Bee-Keepers' Association
B.K.A. Beekeepers' Association
bibliog. bibliography
biogr. biography

C C copy in Mann Library, Cornell University, U.S.A.
c. circa
chap. chapter
cm. centimetres
Co. Company
col. colour/coloured
Cole *The Cole Library of early medicine and zoology* by N. B. Eales, University of Reading, 1969
corr. corrected

D D copy in Trinity College Dublin Library

E ed. edition, edited (editor is spelt out)
eds. editions
enl. enlarged

F ft foot/feet
Fig. Figure
Fraser Dr. H. M. Fraser
Fraser's *History* *History of beekeeping in Britain* (635)
front. frontispiece
Fussell *The old English farming books 1523–1730* and *More old English farming books 1731–1793* by G. E. Fussell, London, Crosby Lockwood, 1947 and 1950

G G copy in Ministry of Agriculture Library, London

H H copy in Hereford Library
H.M. Stationery Office His/Her Majesty's Stationery Office
Henrey *British botanical and horticultural literature before 1800* by B. Henrey. London, OUP, 1975
Hodgson *Children's books on bees and beekeeping* (665)
Hunt *Catalogue of botanical books in the collection of Rachel McMasters Miller Hunt*. 2 vol. compiled by J. Quinby and A. Stevenson. Pittsburgh (PA), Hunt Botanical Library, 1958–61
Hon. Honorary

I illus. illustrated (with figures/diagrams/plates) illustrator is spelt out
Inc. Incorporated
incl. including

J Johansson *Apicultural literature published in Canada and the United States* (see page 22)

L Lisney *A bibliography of British Lepidoptera 1608–1799* by A. A. Lisney. London, Chiswick Press, 1960

M M copy in Miller Library, Madison, U.S.A.
Manley R.O.B. Manley
Moir Scottish Beekeepers' Association Moir Library, Edinburgh
ms manuscript written before 1500 (page 25)
mt metaphorical title (page 235)

N N copy in University of Reading Library
n.d. not dated
No. Number

O O copy in Bodleian Library, Oxford
obs. observation(s)

P P copy in Patent Office Library, London
p. pages (number in book)
pl. plate, plates
Plomer *A dictionary of the printers and booksellers in England, Scotland and Ireland* 4 vol. by H. R. Plomer and others. London, Bibliographical Society, 1977
pp. pages (section referred to in book)
pr. printer, printed
pseud. pseudonym
pub. published

Q q.v. which see

R R copy in Rothamsted Experimental Station Library, Harpenden
R.T.S. Religious Tract Society
ref. reference
refs. references
Rev. Reverend
rev. revised

S S copy in Scottish Beekeepers' Association Moir Library, Edinburgh
S.T.C. *A short title catalogue of books printed in England, Scotland, & Ireland, and of English books printed abroad 1475–1640* by A. W. Pollard and G. R. Redgrave. London, Bibliographical Society, 1926 (now in course of revision)
Sec. Secretary
Soc. Society

U U copy in University of Cambridge Library

V Vol. Volume(s)
vol. volume(s)

W W copy in Library of Congress, Washington, D.C., U.S.A.
W.B.C. William Broughton Carr
Walker *Catalogue of bee books* (814)
Wing *Short-title catalogue of books printed in England, Scotland, Ireland Wales, and British America and of English books printed in other countries 1641–1700* 3 vol. by D. Wing. London, Index Society, 1945

Enrichment of the information provided

Libraries with beekeeping collections

All the libraries whose holdings are listed in the entries have outstanding collections of bee books. They are:

United Kingdom and Irish Republic

A British Library, Department of Printed Books, Great Russell Street, London WC1B 3DG

B International Bee Research Association Library, Hill House, Gerrards Cross, Bucks. SL9 0NR

D Trinity College Library, Dublin 2

G Ministry of Agriculture Library, 3 Whitehall Place, London SW1 2H8

H Hereford and Worcester County Libraries, Hereford Library, Broad Street, Hereford HR4

N University of Reading Library, Whiteknights, Reading RG6 2AG

O Bodleian Library, Oxford OX1 3BG

P Patent Office Library (now part of Science Reference Library, British Library), 25 Southampton Buildings, Chancery Lane, London WC2A 1AW

R Rothamsted Experimental Station Library, Harpenden, Herts. AL5 2JQ

S Scottish Beekeepers' Association Moir Library, Central Public Library, George IV Bridge, Edinburgh EH1 1EG

U University of Cambridge Library, West Road, Cambridge CB3 9DR

U.S.A.

C Mann Library, Cornell University, Ithaca, NY 14853

M Miller Library, Steenbock Library, University of Wisconsin, Madison, WI 53706

W Library of Congress, Washington, D.C.

Other libraries have important collections of these books, but for one reason or another their holdings are not listed, except that a reference to the Cole and Hunt catalogue in an entry implies that the relevant Library holds the book:

United Kingdom

Royal Entomological Society of London, 41 Queen's Gate, London SW7 5HU

Hampshire Beekeepers' Association, c/o Hampshire College of Agriculture, Sparsholt, nr. Winchester, Hants.

Cole Library, University of Reading Library, Whiteknights, Reading RG6 2AG

National Museum of Wales, Cathays Park, Cardiff CF1 3NP

Canada

University of Guelph, Guelph, Ontario N1G 2W1

U.S.A.

National Agricultural Library, U.S.D.A., Beltsville, MD 20705

New York Public Library, Fifth Ave. and 42nd St., New York 18, NY

Jager Library, University of Minnesota Library, St. Paul, MN 55101

Hunt Botanical Library, Carnegie-Mellon University, 5000 Forbes Ave., Pittsburgh, PA 15213

New Zealand

National Beekeepers' Association of N.Z., c/o Box 4048, Wellington

Publications with useful bibliographical information

The books most extensively used in compiling the *Bibliography*, and most frequently quoted in it, are:

Fraser, H. M. (1958) *History of beekeeping in Britain.* London, Bee Research Association. 106 pp.

Scottish Beekeepers' Association (1950) *Catalogue of the Moir Library.* Edinburgh, Scottish Beekeepers' Association, 193 pp. (Supplement 1950–1962, 97 pp.).

Walker, H. J. O. (1929) *Catalogue of bee books, collected and offered for sale.* Budleigh Salterton (Devon), published by the Author, 144 pp.

Johansson, T. S. K. & Johansson, M. P. (1972) *Apicultural literature published in Canada and the United States.* New York, published by the author (sold by I.B.R.A.), 102 pp.

Hodgson, N. (1973) *Children's books on bees and beekeeping.* London, Bee Research Association, iii, 55 pp.

The following library catalogues are in the I.B.R.A. Library, and all but the last three have been published.

Ministry of Agriculture and Fisheries (1937, 1939) *Cowan Memorial Library: hand list of books in English printed after 1930.* London, 35 pp.

The Charles C. Miller Memorial Apicultural Library catalogue (1936). Madison (Wisconsin), University of Wisconsin, 283 pp.

British Bee-Keepers' Association (1912) *Catalogue of the Library of British Bee-Keepers' Association.* London, 29 pp.

Johansson, T. S. K. & Johansson, M. P. (1972) *Bee-library of the late Rev. L. L. Langstroth.* Bee World 53(1): 22–27

Rothamsted Experimental Station (1940) *Library catalogue of printed books and pamphlets on agriculture published between 1471 and 1840.* 2nd ed.

Ministry of Agriculture (1950) *Books in the Cotton Collection at the Ministry of Agriculture Library.*
Unpublished typed list, 5 pages, prepared for B.R.A. probably by Mrs. R. M. Duruz. About 50 books in date order, 1609–1875, with an author index.

Jager, F. (1929) Catalogue of Bee Library. *Unpublished manuscript*, 64 + 20 pages

Drory, E. (1903 ?) *Catalogue of bee books in English in the Drory Library, Berlin.* Photocopies of catalogue cards, 68 pages

My bee book by W. C. Cotton (1842), which was reprinted in 1977 (see 238), gives an extensive list of earlier books. The index to the first thirty volumes of *Bee World* (entry 581a) lists all books reviewed in the journal from 1919 to 1949. The years from 1950 onwards are covered by a computer-generated index to the companion journal *Apicultural Abstracts*; the published index (entry 604) continues to 1972. In 1966 *Bee World* published 'Lists of books recommended by the Bee Research Association to candidates for beekeeping examinations' (Vol. 47, pp. 37–41).

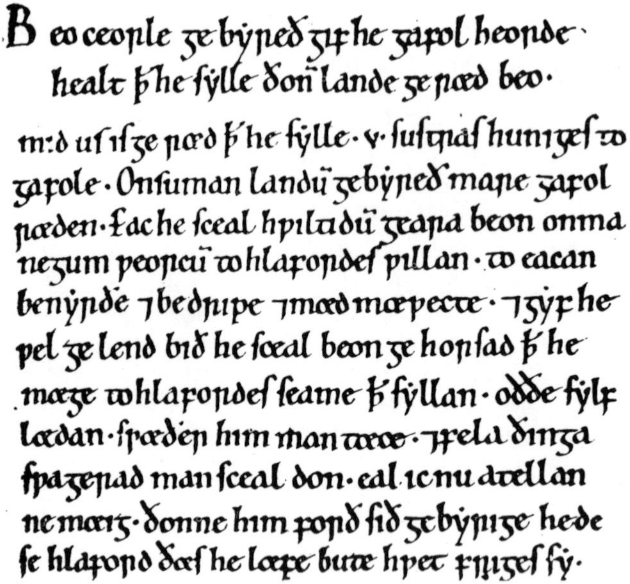

Fig. 1. Passage setting out the rights and duties of the beekeeper in A.D. 1000, from *Rectitudines singularum personarum* (entry *ms* 3). Reproduced by permission of the Master and Fellows of Corpus Christi College, Cambridge. The passage, with its English translation, forms the frontispiece of H. M. Fraser's *History of beekeeping in Britain* (entry 635).

Manuscripts before 1500

The legacy of classical knowledge and beliefs

ms 1 500 and later

Ancient laws of Ireland. A collection supposedly codified by St. Patrick (438–441), but now dated to the 6th/7th centuries and later.

Published in 6 volumes (in Irish and in English translation) under directions of the Commissioners for Publishing the Ancient Laws and Institutes of Ireland. / Dublin, Thom, also London, Longmans and others, 1865–1901. / 24 cm. / A

Vol. 4, 1879, contains the section entitled *Judgments concerning bees* pp. 161–204.

Fraser's *History* pp. 13–15 summarizes the judgments, which deal with ownership of swarms, hive stealing, penalties for stinging by bees (a man's full meal of honey), and other damage committed by them, e.g. blinding a person. There is also information on hives and their placement, and on honey storage.

ms 2 mid-900s

Ancient laws and institutes of Wales. Comprising laws attributed to Hywel Dda (king of Wales in the mid-900s), and surviving in manuscript from the 1200s onwards.

Pub. 1841 under above title, ed. A. Owen. / London, Record Commission. / 1005 p./39 cm. / A

Passages on bees: Vol. I pp. 284, 288, 500, 502, 562, 738, 740, 794; Vol. II pp. 94, 96, 110, 212, 556, 600.

This book is also in the Libraries of the London Record Office and the National Museum of Wales, Cardiff, from which I.B.R.A. has received photocopies of the pages listed and of relevant pages in other published texts.

Like *ms* 1, the Welsh laws were based on much older material, and little affected by Roman influence. References to bees include the values and stocks and swarms at different times of the year, for instance: an old stock 24d, first swarm 16d, bull swarm [stag swarm, second swarm] 12d, third swarm 8d; first swarm from first swarm 12d, from bull swarm 8d, from third swarm 4d. 'If a swarm go out after the calends of August it is a wing swarm. And so they continue until the calends of winter. From the calends of winter onward, each is an old stock and is 24d in value: only a wing swarm does not become an old stock until the calends of May, and then it is 24d in value.'

ms 3 1000

Rectitudines singularum personarum. Setting out the duties and privileges of people who were bound to the land.
See *Ancient laws and institutes of England*, ed. B. Thorpe./London, Record Commission, 1840.
The beekeeper and swineherd belong to the lowest rank, the duties and privileges of the former being as follows: 'The beekeeper, if he has a taxable stock of bees, is to give (as rent) from it whatever is the custom of the area. With us it is the custom for him to pay 5 sesters of honey as rent. In some areas a greater rent is required. Moreover at certain times of the year he is to be employed on various duties at his lord's pleasure, particularly boon-ploughing, boon-reaping and mowing. And if he has good land he is to have a horse which he may provide for the lord's service or else lead himself, whichever is ordered. And [he is to do] various similar duties as is fitting. I am not able to enumerate them all. When he meets with death, his lord is to take charge of what he leaves, except for any freehold property that may be amongst it.'

The passage in the manuscript is reproduced in Fig. 1.

ms 4 12th century

Bestiaries. According to Fraser's *History* p. 27, the bee was not included in *Physiologus*, the chief source of mediaeval bestiaries, which was written between the second and fifth centuries, probably in Egypt. A section on the bee was added in the late bestiaries, most of the material being derived from the *Origines* of St. Isidore of Seville (c. 560–636) and the *Hexameron* of St. Ambrose, Bishop of Milan (c. 340–397).
Cambridge University Library holds the manuscript of a 12th century bestiary (Ii 4 26) which may have been written (copied from earlier material) at the Abbey of Revesby in Lincolnshire. This manuscript was published by the Roxburghe Club under the title *The bestiary*. ed. M. R. James. Oxford, 1928.
A translation by T. H. White was published under the title *The book of beasts* /London, Jonathan Cape, 1954. 296 p. illus./A (B); *see* 826.
In White's book the entry on bees occupies 7 pages which include an illustration showing bees in a skep, and a long quotation from Cutwode, 1599 (*see* 15). 'Bees are called Apes either because they cling to things with their feet (*a pedibus*) or else because they are born without feet, for they only grow their feet and wings later on.' 'Many people have proved that these creatures are born from the corpses of cows.' Bees are commended for their hard work and for their loyalty to their king, and their system of holding everything in common, even procreation: 'For they suddenly produce the greatest swarm of progeny, picking up grubs with the mouth from leaves and grasses.'

NECKHAM, Alexander *ms* 5 c. 1200

De naturis rerum.

Published as part of the *Chronicles and memorials of Great Britain and Ireland during the Middle Ages* under the direction of the Master of the Rolls. Vol. 34./ed. Thomas Wright. London, Longmans, 1863./521 p. 24 cm./A

Cole 77a.

Neckham lived from 1157 to 1217. The date of the book, which was written in Latin, is not certain. It was in part a manual of the scientific knowledge of the time, but contained also Neckham's moralizations and stories. The bee matter forms Chapter 163 and is based on Virgil and St. Isidore.

Neckham's mother was wet nurse to Richard I.

BARTHOLOMAEUS ANGLICUS *ms* 6 c. 1250

De proprietatibus rerum./Written in Latin by an English friar.

Translated into English by John of Trevisa; the translation was finished in 1398, but the earliest printed edition, produced by Wynken de Worde, is dated 1495. 29 books./Bees are in Book XII, the material being taken from St. Ambrose and Aristotle.

Subsequent eds. 1535, 1582./A

ms 7 c. 1250

Ceo est hosebonderie./A book on husbandry written in Norman French by an unknown author.

Included in *Walter of Henley's Husbandry together with* [other works] (transcribed and translated by E. Lamond)./London, Longmans, 1890./A

Each hive should give two swarms. In some places they are fed in the winter.

ms 8 1269–1270

The account-book of Beaulieu Abbey.

Published in 1975 ed. S. F. Hockey./London, Royal Historical Society, 1975./vi, 348 p./B

Camden Fourth Series Vol. 16 (*see* 663a)

Includes much information on the honey and wax produced in 1269 and 1270 in apiaries on the Beaulieu Abbey estates, and some on the hives in use.

ms 9 1299–1490

[*Legal records of the forests.*]

Published in *The royal forests of England.* ed. J. C. Cox./London, Methuen, The Antiquary's Books, 1905./xvi, 372 p. 34 pl./A

Fraser's *History* (p. 24) gives details of a number of the cases that relate to bees.

ROLLE, Richard, of Hampole *ms* 10 c. 1350
[*Various manuscripts.*]
Published in *English writings.*/London, Early English Text Soc., 1866./ Three qualities of the bee pp. 8–9./A
Anr. ed./1921./A
Anr. ed. edited by Hope E. Allen./Oxford, Clarendon Press, 1931. /Bees pp. 54–56./A
Selected works. Translated by G. C. Heseltine./London, 1930./illus. Bees pp. 103–104./A S

The following extracts are from the translation by G. C. Heseltine.

'The bee has three characteristics. One is that she is never idle, and she is against them that will not work, and casts them out and puts them away. Another is that when she flies she takes earth in her feet, that she be not lightly carried up in a puff of wind. The third is that she keeps her wings clean and bright. Thus righteous men that love God are never in idleness . . . They take earth; that is, they hold themselves vile and earthly, that they be not blown about with the wind of vanity and of pride. They keep their wings clean; that is they fulfil the two commandments of charity conscientiously, and they keep other virtues unblended with the filth of sin and unclean lust.'

Richard, son of William Rolle, was born about 1300 at Thornton Dale near Pickering, Yorkshire. His parents were comparatively poor, but

---

Fig. 2. Page giving gardening instructions for January, from *Yn y lhyvyr hwnn*, the first book to be printed in Welsh (about 1546, entry 2a).

In this month prune your [fruit] trees and they will not split. Uproot the shrubs and thorns from your meadow and they will not grow again. And do this particularly within four days of the disappearance of the moon. Dig your garden and spread it with dung, move your bees. Uncover the roots of your fruit trees, especially the old ones which have not borne fruit. Make use of your plough, prepare your land for wheat and rye, plough your fallow land for oats so that it will be soft.

Reproduced by permission of the National Library of Wales, which also supplied the English translation. The same page is reproduced as Fig. 3 of *Y fêl ynys* by M. R. Williams (entry 830).

Januarius.

Mis Jonawr. y sy o .xxxi. o dyddieu,

xix	A	Duw calan.
viii	b	
	c	
xvi	d	
v	e	
	f	Gwyl ystwylh.
xiii	g	
ii	A	
	b	
x	c	
	d	Yr haul yn y dyfrwr.
xviii	e	Gwyl Llwcharn.
vii	f	Gwyl Elen ac erbin. Gwyl hilari.
	g	Kalan chwefrawr.
xv	A	
iiii	b	
	c	
xii	d	
i	e	
	f	
ix	g	
	A	S. Vinsent.
xvii	b	
v	c	
	d	Gwyl bawl. pan droes yr fawn.
xiiii	e	
iii	f	
	g	
vi	A	
xix	b	
	c	Gwyl ewryd sanct.

Yny mis ymma torr dy goed defnydd ag ny holltan. Diwreidda y koedach ac dyrysslli oth wairglod ag ny thyfuan eilwaith. A gwna hynn yn enwedig o ddyuewn pedwar nywarnod y diwed y lheyad. Val dy ardd a theila hi a thomm, lymyd dy wenyn. Diuoetha wreiddeu dy goed ffrwyth, yn enwedig or rhai y vo hen ac heb ddwyn ffrwyth. Gwna defnyddyon dy aradyr. bryn ara dy dir y wenith ath ryg, torr dy wndwn ygeyrch val y bo medal.

A.i. Februa-

with the assistance of Thomas Neville, Archdeacon of Durham, they sent him to Oxford. He left at the age of 18, to become a hermit in the house of John of Dalton, improvising a hermit's habit from two kirtles (skirts) of his sister. When he left, he continued to live a religious life of fasting and prayer, and after some wanderings finally settled down by the nunnery at Hampole.

ms 11 late 1300s

Mum and the sothsegger.
Published as *A poem written in the West Midlands at the end of the 14th century*. London, Early English Text Society, 1936./Original Series no. 199./A (B)

The narrator is searching for a sothsegger (truth-teller) and comes upon an old man killing drones to preserve the honey in his hive. The old man gives an account of life in the bee hive (lines 966–1086). The bee matter is based partly on Bartholomaeus (*ms* 6).

PALLADIUS, Rutilius Taurus Emilianus *ms* 12 c. 1420
Palladius on husbondrie. A verse translation, made (probably by a monk) about the time of Chaucer, of *De re rustica*, 4th century. The manuscript was formerly in Colchester Castle, but has apparently been mislaid or lost.
Published as *Palladius on husbondrie*. ed. Barton Lodge for the Early English Text Society./London, Trubner, 1873./Bees pp. 37–39, 155./A D N O

The fact that the manuscript was translated exemplifies the strong domination of the earliest English writers by Latin authorities, but some of the advice given in the poem was suitable only for Mediterranean countries. One of the stanzas is reprinted in Fraser's *History*, p. 23.
The 15th century translation cannot be fully understood by the modern reader, but E. D. Collins prepared a modern prose rendering of it for I.B.R.A. (E132), and this is held by B.
Barton Lodge stated that the manuscript in Colchester Castle was the only copy, but Fraser believed that there were others.

Chronological bibliography of Books

1500 to 1599

Books written 'out of experience'

[FITZHERBERT, [John] 1 1523
(attributed by some to Sir A. Fitzherbert)
[*Book of husbandry*] Here begynneth a newe tracte or treatyse moost profytable for all husbande men. / London, R. Pynson, [1523]. / A
This book was published in 11 further editions between 1525 and 1598, and was reprinted in *Certain ancient tracts concerning the management of landed property.* / London, C. Bathurst and J. Newbery, 1767.
A G N O R S U. W
S.T.C. 10994–11004. / Hunt 33.

The bee material is contained in just over two pages, and half of it deals with taking swarms. There is a reference to feeding a late swarm, and to reducing the entrance in winter. The 1523 edition mentions drones which will eat the honey and gather nothing. By the 1568 edition, 'now lately corrected', the husbandman is recommended to kill the drones, as there is a saying that she has lost her sting and will not labour as the others do.

Blagrave (41) used Fitzherbert's beekeeping matter without acknowledgement.

2 1526
The grete herball whiche geveth parfyt knowlege and understandyng of all maner of herbes . . . / London, Peter Treveris, 1526. / A
Subsequent eds. 1539, 1561. / A
S.T.C. 13176–79. / Henrey 168–171. / Hunt 25.

Rohde's *The old English herbals* gives a full account of the book, including the passage on honey made by artifice and craft of bees. Bees make honey and wax from flowers. There is tame honey from hives, and wild honey found in woods which is more bitter. Honey is white in cold places and brown in warm places. It ought to be put in medicine, and may be kept 100 years.

This book contains a woodcut which is the first illustration of bees printed in England; see frontispiece. The book was translated from the French *Le grant herbier*; this has no illustration showing bees. According to Rohde the illustrations of *The grete herball* are inferior copies of those in the later editions of the *Herbarius zu Teutsch*. The only copy of the woodcut we have traced is published in *Dē grotē herbari*' printed by Claes die Graue, Antwerp, 1526 (A's copy has shelf mark 546 i 8, catalogued under Cube, Johann von). In the following year Treveris used the same woodcuts to illustrate Braunschweig's *The vertuose boke of distyllacyon of the waters of all maner of herbes* (1527), but although Chapter CXXI deals with hony water (The best hony of bees is that is whyt and that of the bees dwellinge farre from the see and from unclene places), the bee picture is not included.

BARTHOLOMAEUS ANGLICUS 1535
De proprïetatibus rerum. (*See* ms 6.)

2a c.1546
Yn y lhyvyr hwnn. (In this book.)
This, the first book to be printed in Welsh, included gardening instructions month by month. Those for January, which mention bees, are reproduced in Fig. 2, from the copy of the book in the National Library of Wales.

TUSSER, Thomas 3 1557
A hundreth good pointes of husbandrie. / London, R. Tottel, 1557. / A R
Further editions were soon published, and in 1573 the book was expanded into *Five hundred pointes of good husbandrie*, of which there were at least 19 editions up to 1672. The book was reprinted in 1931 and 1954, and the first edition (1557) was reprinted in the English Experience series as No. 628 (1973). / A G O P R S U. C W

S.T.C. 24372–92. / Wing T 3368–69. / Cole (1931 reprint) 245. / Hunt 325, 425.

The original book contained 100 verses of advice on husbandry. Verses 84 and 85 dealt with bees:

Saint Mihel byd bees, to be brent out of strife:
sajnt John bid take honey, with fauour of life.
For one sely cottage, set south good and warme:
take body and goodes, and twise yerely a swarme.

At Christmas take hede, if their hiues be to light:
take honey and water, together wel dight.
That mixed with strawes, in a dish in their hiues:
they drowne not, they fight not, thou sauest their lyues.

In later editions the plan was extended with a monthly abstract followed by a fuller version of the month's husbandry. There were many additional poems and digressions, and in the 1573 edition a verse autobiography was added. The bee matter in the 1580 edition is as follows:

September

20 Drive hive good conie,
for wax and for honie.
No driving of hive,
till yeere's past five.

21 Good dwelling give bee
or hence goes shee.

20 Now burne up the bees that ye mind for to drive
at Midsomer drive them and save them alive:
Place hive in good ayer, set southly and warme,
and take in due season wax, honie and swarme.

21 Set hive on a plank, (not too low by the ground)
where herbe with the flowers may compas it round:
And boordes to defend it from north and north east
from showers and rubbish, from vermin and beast.

December

23 Helpe bees, sweet conie
with licour and honie.

23 Go looke to thy bees, if the hive be too light,
set water and honie, with rosemarie dight.
Which set in a dish ful of sticks in the hive,
from danger of famine yee save them alive.

May

22 Watch bees in May
for swarming away.
Both now and in June
marke maister bees tune.

22 Take heede to thy bees, that are readie to swarme,
the losse thereof now is a crownes worth of harme:
Let skilfull be readie and diligence seene
least being too careles thou loosest thy beene.

[GESNER, Conrad] 4 [1559]

The treasure of Euonymus conteyninge the wonderfull hid secretes of nature . . . translated (with great diligence and laboure) out of Latin by Peter Morwyng felow of Magdaline Colleadge in Oxford./Imprinted at London by John Daie dwelling over Aldersgate beneath Saint Martines, [1559]./10 leaves, 408 p./A

Of aromaticall wynes that is made of spices p. 392. Of swiet wines spiced p. 396. Three waies to make nectar p. 401. Scattered references to honey.

Anr. ed./*A new book of destillatyon of waters called the treasure of Euonymus.*/London, John Day, [1565]./10 leaves, 408 p./8 leaves A Reprinted 1969 English Experience No. 97.

S.T.C. 11800–01/Hunt 50..

BULLEIN, William 5 1562

Bullein's bulwarke of defence againste all sicknes, sornes, and woundes that dooe daily assaulte mankinde . . . gathered and practised from the most worthie learned, both old and newe: to the greate comforte of mankinde: Doen by Willyam Bulleyn and ended this Marche Anno Salutis 1562./London, Iohn Kyngston, 1562./Bees and honey, folios iii–iiii./A

Anr. ed./1579./A O

S.T.C. 4033–34.

The first part of the work, the *Booke of simples*, consisting of a conversation between Marcellus and Hilarius. After hearing Hilarius extol the virtues of honey Marcellus says, 'I did never here any thing so greatly commended as thou haste commended Honey.' There is also mention of wax. Later in the *Booke of compoundes* various recipes are given using, of course, honey and wax as well as other simples.

MAPLET, John 6 1567

A greene forest, or, A naturell historie, wherein may be seene . . . lastly of brute beastes, foules, fishes, creeping wormes and serpents, and that alphabetically, so that a table shall not neede./London, H. Denham, 1567./8 leaves, 111 folios, 1 leaf. 14 cm./Yewe tree bad for bees p. 65. Of the bee p. 72./A

Anr. ed./1614./A Reprinted London, Hesperides Press, 1930.

S.T.C. 17296./Cole (reprint) 272

HILL (or HYLL), Thomas, Londoner 7 1568

The proffitable arte of gardening, now the third tyme set fourth: to whiche is added muche necessary matter, and a number of secrettes . . . to this annexed two proper treatises, the one entituled the marveilous government, propertie, and benefite of the bees, with the rare secrets of the honny and waxe . . . Englished by Thomas Hill, Londiner./London, Thomas Marshe, 1568.

The bee treatise (see below) has separate pagination, but the register is continuous. There is a portrait of Hill and a dedication to the Worshipfull Mayster M, Gentilman.

A pleasaunt instruction of the parfit orderinge of bees, with the marveilous nature, propertie & governement of theim: and the myraculous uses, bothe of their honny and waxe (serving diversly) as well in inwarde as outward causes: gathered out of the beste writers. To which is annexed . . . certaine husbandly coniectures of dearth and plentie for ever . . . Englished by Thomas Hyll, Londyner./1568./6 leaves, 83 p. 14 cm./Bees to p. 51./A U

Anr. ed./London, Thomas Marshe, 1572./14 cm./A (gardening only). W
Anr. ed./London, Henrie Bynneman, 1574./The bee treatise now called *A profitable instruction of the perfite ordering of bees* . . . etc./4 leaves, 87 p./This and subsequent eds. 18 cm./Bees to p. 44./A O R. M
Anr. ed. There is now added *The booke of the arte or craft of planting and graffing.*/London, Henrie Bynneman, 1579./The bee treatise 4 leaves, 92 p. (pagination rather irregular)./Bees pp. 1–48, 41–43./A D G O. C M W
Anr. ed./London, Robert Walde-graue, 1586./The bee treatise 4 leaves, 92 p./Bees pp. 1–48, 41–43./A O R
Anr. ed./London, Edward Allde, 1593./The bee treatise 4 leaves, 92 p./Bees to p. 46./A O S
Anr. ed. *The arte of gardening* . . ./London, Edward Allde, 1608./The bee treatise printed by H[enry] B[allard], 1608./4 leaves, 92 p./Bees to p. 46/A B (bees only) O P U. C M

S.T.C. 13491–97./Cole (1593 ed.) 304./Henrey 195–201./Hunt 102, 117, 167, 185.

Examination of all A's copies shows that the bee treatise was always intended to be annexed to *The arte of gardening*, although copies exist of the bee treatise on its own. The first known edition, 1563, has no bee matter. Hill's work is based without acknowledgement on a Latin work, *Pantoplion* by G. Pictorius, a German physician (1563), although the word 'Englished' acknowledges the fact that Hill was a translator. However, the 1574 edition states that the work was 'by' Hill, and it may have been this fact which caused Charles Butler to complain so vigorously (18): 'Whom one T.H. of London translating word for word into English as wel as he could, concealing the authors name, adventured to publish in his owne name.'

There is a more complimentary reference to Hill in Baker's translation of Gesner's *Newe jewell of health*, 1576. Baker, who was Queen Elizabeth's 'Chirurgian', names a number of those who have prepared medicines for his patients, 'and also one Thomas Hyll, who for his excellent knowledge in this Arte is not to be left out, who dyd also take paynes in this worke, but before it coulde be brought to perfection, God tooke him to his mercie.' Translator or not, Hill deserves the credit for producing what is undoubtedly the first British bee book. From a brief examination it appears that the bee matter is the same in all editions.

Hill's later work *The gardeners labyrinth*, 1592, has a picture of 'The maner of watering with a pumpe in a tubbe' showing bees and two skeps in the corner of the garden. This picture is reproduced in Rohde's *Old English gardening books*.

MASCALL, Leonard 8 [1570]
Diary.
Published as *The countryman's jewel*. ed. Marcus Woodward./London, Chapman Hall, 1934./*Of the hony-bee* . . . pp. 114–124./(B)
Mascall was the author of books on trees, 1572; poultry, 1581; cattle, 1596.

HERESBACH, Conrad 9 1577
Foure bookes of husbandrie, collected by M. Conradus Heresbachius, Counsellor to the hygh and mightie Prince, the Duke of Cleue: conteyning the whole arte and trade of husbandrie. . . . Newely Englished and increased by Barnabe Googe, Esquire. [From *Rei rustica*, Cologne, 1573.]/London, R. Watkins, 1577./Entreating of bees 4th book, leaves 173–193./A O R. W

Anr. ed./London, John Wight, 1578./A R
Anr. ed/London, T. Este, 1586./A O R. M
Anr. ed./London, T. Wight, 1596./O R
Anr. ed./London, T. Wight, 1601./A O
Anr. ed./*The whole art and trade of husbandry* enl. by B. Googe./London, printed by T.S. for Richard More, 1614/Bees, leaves 164–183./A N O S U
Anr. ed. enl. by Gervase Markham./*The whole art of husbandry*. Now renewed, corrected, enlarged and adorned with all the experiments and practices of our English Nation./London, printed by T. Coates for R. More, 1631./A
Anr. ed./*The perfect husbandman* . . . by C. H[eresbach], B.C. [i.e. B. Googe] and C.M. [i.e. G. Markham], ingenious artists./London, T. Bassett, 1658./Entreating of bees pp. 355–385./A
Reprinted 1971, English Experience No. 323.

S.T.C. 13196–02./Hunt 133, 178, 198.

This book, which is in the form of dialogues, has sections on orchardes, wooddes, cattell, poultry, fish, bees, etc., with rather little on bees, mainly from Virgil, Columella etc. It is one of the books discussed by Levett (24).

DODOENS, Ramslittle 10 1578

A niewe herball, or Historie of plantes translated by Henry Lyte./London, Gerard Dewes, 1578./12 leaves, 779 p. 13 leaves. 30 cm./A O. W

Various references to bees. Mint and rue for bee stings pp. 246, 261. 'If a man put Baume (balm) into Bee hyves, or else if the Hyves be rubbed therewithal, it keepeth Bees togither, and causeth other Bees to resorte to their companie' p. 260. Smoke of Nigella (Melanthium) killeth Flies Bees and Wasps p. 279, etc.

Anr. ed./1586./A

Anr. ed. corr./London, E. Bollifant, 1595./A O R

Anr. ed. *A briefe epitome of the new herbal.*/London, Simon Stafford, 1606./A

Anr. ed. corr./London, E. Griffon, 1619./xxii, 564 p. 15 leaves./A O

S.T.C. 6984–88./Henrey 110–114./Hunt 132, 171.

Dodoens' *Cruydtboeck* was translated into French by Charles de l'Escluse and afterwards into English by Lyte, an antiquarian and a plant collector of Lyte's Cary, Somerset. He dedicated his book to Queen Elizabeth I. (*See* Aubrey, 63.) The 1606 edition, which purported to be an epitome of the *Herbal* collected and abridged by William Ram, Gent., does contain some matter abridged from Lyte's translation, but 'it is in reality a compilation of recipes unworthy of the great name it bears' (Rohde, *The old English herbals*).

RABOTENU, Isaac *pseud.*
(MARNIX, Philips van) 11 1579

Beehive of the Romish Church, wherein the author, a zealous Protestant under the person of a superstitious Papist, doth so driely repell the grose opinions of Popery, and so divinely defend the articles of Christianity, that . . . there is not a booke to be founde . . . sweeter for thy comforte, by Isaac Rabotenu. Translated out of the Dutch into Englische by George Gilpin the elder./London, T. Dawson for J. Stell, 1579./346 p./A

Other eds. 1579, 1580./A O. 1598, 1621, 1623, 1636./A O

S.T.C. 17445–59.

The Chapter headings read as if for a book on bees, but the book is a bitter satire on the Roman church. John Stell, in his preface to the English edition, says the book 'containeth good store of wholesome honey. . . . Gentle Reader, thou hast such a book as will make thee privie to all the practices of the Babylonicall Beast (Rome, I mean), the denne of Dragons and Devils'. The frontispiece shows the papal tiara in the shape of a tall skep, from which the Pope—as the king bee—surveys the other bees, who with shorn heads, mitres and cardinals' hats are flying round him, hearing confessions, burying the dead, and saying mass.

VIRGIL (Publius Vergilius Maro) 11a 1589

The bucoliks of Publius Virgilius Maro . . . otherwise called his pastoralls, or shepeherds meetings. Together with his georgiks or ruralls, otherwise called his husbandrie . . . newly translated into English verse by A[b] F[leming]. / London, T.O. for Thomas Woodcocke, 1589. / 4 leaves, 77 p. / A O U

S.T.C. 24817.

The Fourth *Georgic* gives Virgil's account of beekeeping, in verse, which is summarized in Fraser's *Beekeeping in antiquity* (633). This is the earliest English translation we know of; others listed are: 22, 31, 68, 79, 93, 153, 268, 319, 323, 347, 359, 808. We have omitted still other English translations, and all Latin editions.

SOUTHERNE, Edmund 12 1593

A treatise concerning the right use and ordering of bees: newlie made and set forth, according to the author's own experience. (which by any heretofore hath not been done). Better late than never. / London, Thomas Orwin for Thomas Woodcocke dwelling in Paules Churchyard at the signe of the blacke Beare, 1593. / 17 p. / A H R S. M

Mimeographed copy issued by the Scottish Beekeepers' Association, Edinburgh, 1934, with photostat of original title page. / 27 cm. / B. C W

S.T.C. 22942.

This is the first original book in English entirely about beekeeping. Nothing is known of the author, but he speaks of himself as a practical beekeeper. The dedication is to 'The Right Worshipfull Mistres Margaret Astley, wife to John Astley Esquier, Master and Treasurer of her Majesties Jewels and Plate, and Gentleman of her Highnesse privie Chamber.' The books ends with the story of a parson who demanded of a beekeeper his tenth swarm as a tithe. The beekeeper complied, but delivered the swarm by knocking it out upon the ground so that the parson and his family were stung. This story is reprinted in Cotton's *My bee book* (238).

PLATTE, Sir Hugh 13 1594

The jewell house of art and nature. Containing divers rare and profitable inventions together with the sundry new experiments in the art of husbandry, distillation and moulding. Faithfully and familiarly set downe, according to the Authors owne experience, by Hugh Platte, of Lincolnes Inne, Gentleman. / London, printed by Peter Shory dwelling in Breadstreet hill at the signe of the Star, and are to be sold in Paules Churchyard, 1594. / A O R U

Anr. ed. / 1653. / A N O R

S.T.C. 19991./Wing P2390–91./Hunt 261.

Various references to making honey wines and spirits, also 'How to steale bees' p. 66.

GERARD, John 14 1597

The herball, or Generall historie of plantes gathered by John Gerarde of London, Master in Chirurgerie./Imprinted at London by Iohn Norton, 1597./illus. 32 cm./A D O R . W

Anr. ed. enl./London, T. Johnson, 1633./illus. c. 31 cm./A G N O R U. W

Anr. ed./1636./A O R. This ed. is very similar, but not identical, to the 1633 ed.

In 1975 Dover Books published a facsimile reprint of the 1633 ed. with a few pages taken from the 1636 ed.

S.T.C. 11750–25./Cole (1633 ed.) 358./Henrey 154–156./Hunt 175, 223, 230.

The 1597 edition on p. 1385 deals with fusse bals 'with which in some places of England they use to kill or smolder their Bees, when they woulde drive the Hives, and bereave the poore Bees of their meate, houses, and lives.' Gerard's *Herball* largely consisted of a translation of Dodoens' *Pemptades*, but this passage is not in the Latin original and must be one of Gerard's own comments. It may be the earliest reference to puff ball for stupefying bees—see also Cutwode (15). There are some other references to bees—rubbing hives with balme p. 560, and to the following plants recommended as remedies for bee stings; mallow, mint, rue, bay or laurel.

Some modern versions of the *Herball* omit the bee matter.

CUTWODE, T. 15 1599

Caltha poetarum: or The bumble bee./London, Thomas Creede for Richard Olive, 1599./38 leaves. 15 cm./A (B)

Reprint for Roxburghe Club./London, J. & B. Bensley, 1815./A

S.T.C. 6151.

This is a long poem telling how the bee was accidentally shot by Cupid and fell in love with a Marygold whom he called Lady Caltha. Cutwode did not seem to know the difference between bumble bees and hive bees. In the poem the bee, who is male throughout, loses his sting and becomes a drone. Verse 112 refers to 'spungie Paddocks and the foule fusball.'

1600 to 1699

The beginnings of bee science

STEVENS (ESTIENNE), Charles
& LIEBAULT, Jean 16 1600

Maison rustique, or, The countrie farme, compiled in the French tongue by Charles Stevens and John Liebault, Doctors of Physiche. Translated into English by Richard Surflet, Practitioner in Physiche. / London, E. Bollifant for B. Norton, 1600. / Bees pp. 401–421. / A O R. W

Anr. ed. / London, A. Hatfield for J. Norton and John Bill, 1606. / A O R
Anr. ed. rev. & enl. by Gervase Markham. / London, Adam Islip for John Bill, 1616. / Bees pp. 316–332. / A O R S U. M W

S.T.C. 10547–49. / Hunt 202.

Estienne wrote the *Praedium rusticum* (Paris, 1554) in Latin. After his death in 1564 it was revised and published in French by Liebault (Paris 1570 and 1572) as *L'agriculture et maison rustique*. Surflet made a faithful English translation, and Markham did not alter the bee material much. There are long excerpts from *Maison rustique* in *The countryman's jewel* (8).

PLINY (the elder, **Caius Plinius Secundus**) 16a 1601

The historie of the world, commonly called the natural historie of C. Plinius Secundus. Translated into English by Philemon Holland. / London, Islip, 1601. / 2 vol. / A

2nd ed. / London, Islip, 1634. / 2 vol. / A

S.T.C. 20029–30a.

Many references to bees; *see* Fraser (633).

TOPSELL, Edward 17 1607

Historie of four-footid beastes collected out of the writings of Conradus Gesner. / London, Jaggard, 1607.
This contains no bee material but was followed shortly by, and is often bound with:
Historie of serpents. / London, Jaggard, 1608. / 5 leaves, 315 p. 4 leaves, illus. 30 cm. / Bees pp. 64–82, written by Dr. Bonham. / A O S

2nd ed. (including both the above works) Whereunto is now added the

Theater of insects . . . bees, flies, caterpillars, spiders, worms etc. A most elaborate work: By T. Muffet, Dr. of Physick. The whole rev. and enl. by J. R[owland], M.D./London, E. Cotes for G. Sawbridge, 1658./7 leaves, 1130 p./Bees pp. 637–650, 3 leaves./illus. (*Historie of serpents*), and 889–921 with skep on title page (*Theater of insects*)./A (B) D O R S. C M

Reprinted 1973 in the English Experience series No. 561 (beasts 1607) and 562 (serpents 1608).

The 1658 edition was reprinted with an introduction by Willy Ley in 1967 by the Da Capo Press in the U.S.A. and by Frank Cass & Co. in the U.K.

S.T.C. 24123–24./Cole 343./Lisney 1, 2.

This book, according to Dr. Casey Wood is 'the great picture book of the 17th century and is nearly always tattered, dog eared and imperfect.' *See also* Mouffet (25); his name is variously spelt, being Muffet on the title page of the whole work in the 1658 ed. but Mouffet on the title page of the *Theater of insects* itself.

BUTLER, [Rev.] Charles 18 1609

The feminine monarchie, or A treatise concerning bees, and the due ordering of them: wherein the truth, found out by experience and diligent observation, discovereth the idle and fond conceipts which many have written anent this subject./Oxford, printed by Joseph Barnes, 1609./12 leaves, 108 leaves./A G O S. M

Anr. ed. *The feminine monarchie: or The historie of bees;* shewing their admirable nature, and properties, their generation, and colonies, their government, loyaltie, art, industrie, enemies, warres, magnanimitie, &c. Together with the right ordering of them from time to time, and the sweet profit arising thereof. Written out of experience./London, Pr. by John Haviland for Roger Jackson, and are to be sold at his shop in Fleetstreet, over against the Conduit, 1623./97 leaves, engraved front./A O R S. M W

Anr. ed. *The feminin' monarchi', or, The histori of bee's.* Shewing etc . . ./ Oxford, printed by William Turner, for the Author, 1634./7 leaves, 180 p. wrongly numbered 182 (there is no p. 113–114), incl. engraved front. Written in the phonetic spelling devised by Butler./A B G N O P R U. C M W

Anr. ed. Translated into Latin by R. Richardson as *Monarchia foeminina.*/ London, 1673./9 leaves, 200 p. front. 15 cm./A B G O S U. C M W

Anr. ed. *Monarchia foeminarum, sive apum historia.*/Oxon, 1682./O

Anr. ed. *The feminine monarchy;* . . . translated into English by W.S./ London, A. Baldwin, 1704./[ii], 150 p. front. 15 cm./A B G O. M

Anr. ed. Printed from the same type as the preceding./London, for G. Conyers, 1704./R. C M

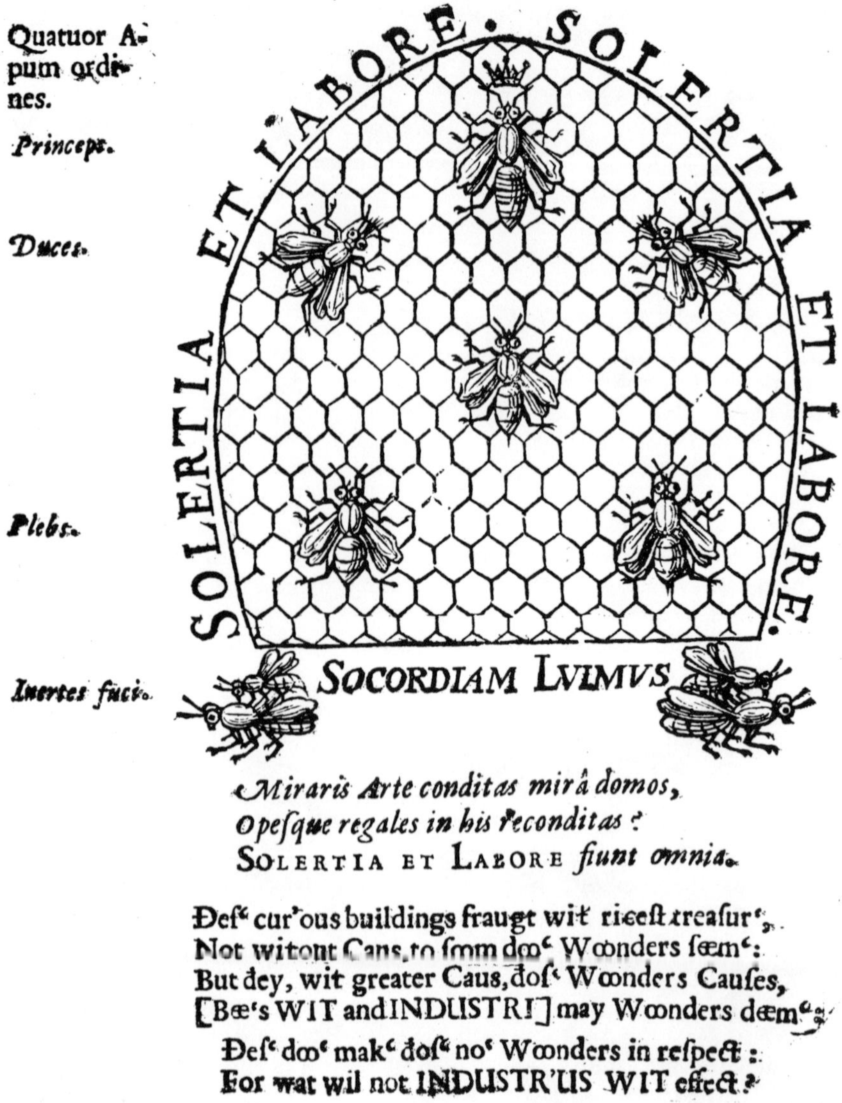

Fig. 3. Illustration first used in the 1623 (second) edition of the Rev. Charles Butler's famous book *The feminine monarchie* (entry 18). The crown above the queen is perhaps in honour of the consort of King Charles II, Queen Henrietta Maria, to whom this edition is dedicated.

1st ed. reprinted 1969 English Experience No. 81.

S.T.C. 4192–94./Wing B6262–63.

This is the greatest early British bee book and contains the best account of skep beekeeping which is available today. Butler (1571–1647) was one of the many clergymen who wrote about bees, and his first edition contains an exhortation to his readers, now that they have had the benefit of his instruction, to pay their tithes justly.

Butler also wrote a book in 1626 on the *Principles of musik*, and in the first edition of his bee book he attempted to describe the piping of the queen at swarming time in musical notation. In the 1623 edition this had been expanded into a four-part madrigal printed inversely on the upper and lower halves of two opposite pages, so that the singers facing each other two and two with the book held between them could each sing his part. Butler was also interested in English grammar, and in his book on that subject in 1633 he advocated a system of phonetic spelling, which he used for the third edition of his bee book.

A leaflet by F. W. Money published in 1953 gives details of Butler's life. See also the *Dictionary of national biography*. Butler was vicar of Wootton St. Lawrence near Basingstoke, where a window in the church is dedicated to his memory; it is illustrated in *The bee book* by Daphne More (731).

The reverse of the 1623 title page is shown in Fig. 3.

[MARKHAM, Gervase] 19 1614

Cheape and good husbandry for the well ordering of all beasts and fowles . . . together with the use and profit of bees./London, T.S[nodham] for R. Jackson, 1614./16 leaves, 162 p./Bees pp. 151–157./A D O

Many eds. to the 15th in 1695./A (B) G O R S. C M W

The 3rd and all subsequent eds. were part of *A way to get wealth.*

1st ed. reprinted 1969 English Experience No. 139.

S.T.C. 17336–40./Wing M611–618./Hunt 259.

F. N. L. Poynter's *A bibliography of Gervase Markham* (Oxford Bibliographical Society, 1962) makes it plain that Markham (1568–1637) was far from being a mere hack writer as has been generally supposed. 'The portrait emerges of a talented man with the kind of insatiable curiosity characteristic of the founders of the Royal Society later in the seventeenth century and resembling them also in becoming a master of simple prose as a medium for communicating his knowledge.' Markham wrote poetry, plays, a novel, and works on horsemanship and military matters, as well as books on agriculture and gardening. In 1623 the publisher Roger Jackson brought together a number of Markham's works including the 3rd edition of *Cheape and good husbandry*, and Lawson's *A new orchard and garden* (20), under the title '*A way to get wealth:* containing the foure principal offices which support and maintain a familie'. This compilation reappeared numerous times with various additional contents until the issue of 1683:

A way to get wealth containing six principal vocations or callings, in which every good husband or house-wive may lawfully imploy themselves . . . the fourteenth time corrected and augmented by the author./ London, Printed by T.B. for Hannah Sawbridge, at the Bible on Ludgate Hill 1683./

This contains the 14th ed. of *Cheape and good husbandry* and the 13th ed. of Lawson's *A new orchard and garden.* There was a final edition—the 15th time corrected . . . printed for A. & J. Churchill, 1695—with the same contents. Poynter gives further bibliographical details. He suggests that *A way to get wealth* was always intended to be published as a whole, although the various parts had different title pages and now often appear separately. As far as we can discover the various editions of both works—and certainly of the bee chapter—are the same. For a much later 18th century re-issue see *The country gentleman's companion* (107).

Markham also revised Estienne's *Maison rustique* (*see* 16), and *The whole art of husbandry*, a reprint of B. Googe's translation of Heresbach 'corrected, enlarged and adorned . . . by G. Markham.' 1631 (*see* 9).

Markham wrote a recommendation in Levett's book (24) in which he described himself as being 'once a great Master of these creatures; neither have I spared to bestow my pen in the advancement of the Art and knowledge of the Bee.' However a comparison between the bee matter in *Cheape and good husbandry* and Lawson's *A new orchard and garden* (20) shows Lawson to be the better practical beekeeper.

LAWSON, William 20 1618

A new orchard and garden . . . with *The country housewife's garden* for hearbes of common use . . . as also the husbandry of bees, with their several uses and annoyances./London, Pr. by B. Alsop for R. Jackson, 1618./illus./A O R

The country housewife's garden has its own title page dated 1617, but both copies we have examined (held by A and R) were clearly issued with *A new orchard and garden*, because the sections are continuous.

About 13 eds. to 1695./A (B) G N O P R S U. C M W

Reprint of 3rd ed. 1626 with preface by E. S. Rohde./London, Cresset Press, 1927./Bees pp. 87–94./A H S U. M

S.T.C. 15329–31./Wing L730–738./Henrey 226–228./Hunt 209, 258, 307.

There is a brief mention of bees on p. 58 of *A new orchard and garden*:

'for cleanly and innocent Bees, of all other things, love and become, and thrive in an Orchard', and there is also a reference to bee boles, but bees are more extensively dealt with in Chapter X of *The country housewife's garden* pp. 19–25.

The illustration is a charming woodcut of straw hives under a penthouse roof and is reproduced facing p. 28 in Fraser's *History* (633).

See also the note to Gervase Markham's book (19).

VENNER, Tobias 21 1620
Via recta ad vitam longam, or, A plaine philosophical discourse of the nature . . . and effects of all such things as by way of nourishments and dieteticall observations make for the preservation of health . . ./London, E. Griffin for R. More, 1620./Hony pp. 99–100. Meathe p. 39./A
Eds. 1622, 1628, 1638, 1650, 1660./A

S.T.C. 24643–48.

Some editions contain a treatise of tobacco and some a treatise of the bathes of Bathe.

VIRGIL (Publius Vergilius Maro) 22 1620
Virgil's eclogues with his booke De apibus: concerning the government and ordering of bees, translated grammatically and also according to the proprietie of our English tongue by J. Brinsley./London, R. Field for T. Man, 1620./A O

Anr. ed London, T. Man, 1633./A O

S.T.C. 24818–89.

BACON, Francis (Viscount Verulam) 23 1627
Sylva sylvarum./London, W. Lee, 1627./A O

P. 225: 'Experiment solitary touching honey and sugar . . . sugar hath put down the use of honey'. Also on meads and on the use of sugar with honey to make a more digestible mead.

Many eds. to 11th./1685./

S.T.C. 1168–73./Wing B327–336.

LEVETT, John 24 1634
The ordering of bees: or, The true history of managing them from time to time, with their hony and waxe, shewing their nature and breed. As also what trees, plants and hearbs are good for them, and namely what are hurtfull: together with the extraordinary profit arising from them. Set forth in a dialogue resolving all doubts whatsoever. By the late unparalell'd experience of John Levett, Gent./London, T. Harper for J. Harison, 1634./12 leaves, 71 p. front./A G O P R S. C M W

Reprinted 1971 English Experience No. 398.

S.T.C. 15555.

Purchas wrote two poems for the book and says in his book (35, p. 109) that the work was written c. 1600. It is a dialogue between Tortona and Petralba, in a style based on a book by Agostino Gallo, Venice 1569. Petralba has inherited some hives and some bee books and he asks Tortona's advice about them. The books chiefly discussed are Southerne's (12) and Googe's translation of Heresbach (9). The frontispiece is a picture of a skep also used for the Latin version of Mouffet's work published the same year (25). The book was published by Levett's son.

MOUFFET, Thomas 25 1634
Insectorum sive minimorum animalium theatrum./London, Cotes, 1634./xx, 326 p. 2 leaves, illus./In Latin./A D O U

2nd ed. translated into English as *The theater of insects* see Topsell (17).

S.T.C. 17993./Lisney 3–6 gives details of 3 issues of 1st ed.

Some authorities believe that Mouffet's daughter Patience was the original Little Miss Muffet. Her father was certainly very interested in spiders.

The book was based on notes by Conrad Gesner which were entrusted to Thomas Penny. Mouffet added new material from Edward Wotton, but publication was delayed by Mouffet's death in 1604. Finally Sir Theodore Mayerne, Court Physician, purchased the manuscript and published the book with a new dedication in 1634. The only English translation of Mouffet's work appeared in the 2nd ed. of Topsell's *Historie* (17); Topsell's first ed. did not contain Mouffet's work.

REMNANT, Richard 26 1637
A discourse or historie of bees, shewing their nature and usage, and the great profit of them. Whereunto is added the causes and cure of blasted wheat, and some remedies for blasted hops and rie and fruit./London, printed by Robert Young for Thomas Slater dwelling in Duck Lane at the White Swan, 1637./2 leaves, 47 p./A R S. M

S.T.C. 20879.

Fraser's *History* describes this as a book ahead of its time; for instance Remnant knew that the workers were female.

PARKINSON, John 27 1640
Theatrum botanicum, the theater of plants: or An herball of a large extent./London, T. Coates, 1640./Bees pp. 6, 44, 308, 456, 715, 733, etc./Many short references to bees./A D O R U. W

S.T.C. 19302./Henrey 286.

Parkinson was a celebrated herbalist, and apothecary to James I and Charles I. His immense book was, with Gerard's *Herball* (14), the chief authority of the day on botany.

BEST, Henry 28 [1641]

Rural economy in Yorkshire in 1641, being the farming and account books of H.B. of Elmswell in the E. Riding . . . edited by Charles Best Robinson./Durham, for the Surtees Society by Geo. Andrews, 1857./Of bees and how to order them pp. 61–69, 107./A (B) N O R S U. M

Published by the Surtees Society as Vol. 33 in 1857 but not previously. For pictures of Best's farm and some account of the book, see article by F. Austin Hyde, *Yorkshire Illustrated* July 1953.

DAY (or DAYE), John 29 1641

The parliament of bees, with their proper characters, in a bee-hive furnisht with twelve honeycombs, as pleasant as profitable./London, for W. Lee, 1641./illus./A D O. M

Anr. ed. in *Works* with facsimile of 1641 ed./1881.
Anr. ed. in *Nero and other plays* ed. by A. Symons./1888./pp. 210–267./A
Anr. ed. in *Nero and other plays* ed. by A. Symons./1904./pp. 210–267./A

Wing D466.

This is a political satire in verse, written as a play and apparently published in 1607, although the earliest known copies are those of 1641. Some lines are quoted by Leigh Hunt in *A jar of honey from Mount Hybla* (262). The following brief account is from *The Oxford companion to English literature:* 'It consists first of the opening of the parliament, with the viceroy's, "Mr. Bee's", opening address; then of a series of "characters" of different bees with their virtues and vices: the hospitable bee, the reveller, the neglected soldier, the neglected poet, the "usuring" bee, the quack-salver, the thrifty bee, the passionate bee. Finally we have Oberon's Starchamber, with the pronouncement of penalties on offenders, the wasp, the drone, and the humble-bee.'

ARM, F. 30 [1646–58]

A treatise of bees: a manuscript as yet unpublished; gives an account of the authors experiences in the Civil War including the destruction of his fifty rare stocks when the Scots army came first before Newcastle in 1642./The manuscript also includes a report of experiments by 'the Bee-master of Oxford' as related to him by that 'ingenious person and mathematician' Dr. Wilkins (*see* Hartlib, 33).

VIRGIL (Publius Vergilius Maro) 31 1649

Works. Translated by John Ogilby./London, Crook, 1649./Bees pp. 95–112 in Fourth *Georgic*./A O

Eds. in 1654 and 1668./A D./Many others.

Wing V608–615.

HARTLIB, Samuel 32 1651

Samuel Hartlib his legacie, or, An enlargement of the discourse of husbandry used in Brabant and Flanders (in various letters addressed to S.H. and by him collected and edited)./London, printed by H. Hills for Richard Wodenothe at the Star under St. Peter's Church in Cornhill, 1651./4 leaves, 131 p./Bees pp. 63–69./A R. M

2nd ed. enl./London, R. and W. Leybourn for R. Wodenothe, 1652. /Bees pp. 49–52./A N O P R S. C
3rd ed./London, by J.N. for R. Wodenothe, 1655./A G O R U. W
Anr. ed. Reissue of 2nd ed./1659./O

Wing H989–991./Hunt 264.

Despite the impression given by the title page, this book does not contain Richard Weston's *Discours of husbandrie used in Brabant and Flanders* but Robert Child's *Large letter concerning the defects and and remedies of English husbandry*. Part 13 (5 pages): 'It's a great Deficiency here in England without question, that we have no more Bees . . .'

HARTLIB, Samuel 33 1655

The reformed commonwealth of bees, presented in several letters and observations to Samuel Hartlib Esq. With the reformed Virginian silkworm. Containing many excellent and choice secrets, experiments and discoveries for attaining of national and private profits and riches./London, printed by John Street for Giles Calvert at the Black-Spread-Eagle at the West-end of Pauls, 1655./2 leaves, 62 p. 1 leaf 40 p. (Silkworm) illus. bibliog./A G O R S. M

Wing H997.

Hartlib's father was a Polish merchant and his mother the daughter of an English merchant. He came to England in 1623 and became a friend of Milton and other Puritans. He wrote on agriculture and on bees, though he was not a beekeeper. His *Reformed commonwealth* gives the earliest known description of the octagonal three-storied hive, with a drawing of it by Christopher Wren. The hive was designed by Mr. Mew, Vicar of Eastington, Glos., who gave one to Dr. Wilkins of Wadham College, Oxford. It was seen and admired by Wren and John Evelyn. It was, in effect, the same hive that Gedde patented in 1675. The Scottish

Stewarton (early 1800s) is very similar, though there is no proof that Robert Kerr knew anything about the earlier octagonal hive. Fraser states: 'My belief is that the cause of the failure of Mew's, Gedde's, Warder's and similar hives (which were intended to be placed in bee houses) is that the hives were built of thin wood. The Stewarton hive was successful because the wood of which it was built was thicker.' Wren's drawing is reproduced in Fraser's *History*, facing p. 31.

Several bee books have been dedicated to queens, but this book, produced during the Protectorate of Oliver Cromwell, was understandably called *The reformed commonwealth of bees*. The final leaf is a list of 'Some writers of bees extant in English' which mentions Southerne, Butler, Levett, Remnant and 'a book promised by Henry Gurnay, gentleman' which was never published.

EVELYN, John 34 [c. 1655]
Elysium Britannicum or The Royal Gardens in three books. An unpublished manuscript of over 900 pages of which the first 342 survive and are in the library of Christ Church, Oxford.
The bee portion is published: *John Evelyn's manuscript on bees from Elysium Britannicum* (774) edited by D. A. Smith. / London, Bee Research Association, 1966. / 40 p. illus. / A B
The main part of the manuscript was written in the 1650s, although Evelyn himself stated that he worked on the book almost 40 years; *see* 774.

PURCHAS, Samuel (the younger) 35 1657
A theatre of politicall flying-insects, wherein especially the nature, the worth, the work, the wonder and the manner of right-ordering of the bee, is discovered and described . . . and in a second part are annexed meditations, and observations theological and moral in three centuries upon that subject. / London, R.I. for Thomas Parkhurst, to be sold at his shop, at the Three Crowns in Cheapside, over against the Great Conduit, 1657. / 13 leaves 387 p. (187 leaves wrongly numbered). / A B G O P R S. C M W
Wing P4224.
The second part was also issued separately as *Spiritual honey from natural hives*. / London, [n.d.]. / R S
This was reprinted by Bagster (222).
Purchas was the son of the author of *Hakluytus Posthumus or Purchas his Pilgrimes*, 1619. He became rector of Sutton in Essex. His book is dedicated to Robert, Earle of Warwick.

STEVENSON, Matthew 36 1661

The twelve moneths . . . or . . . Ordering of cattle and bees./London, by M.S. for T. Jenner, 1661./Scattered notes on bees./A
Wing S5510.

PEPYS, Samuel, F.R.S. 37 1663, 1665

Diary entry for 11 December 1663

I to the Coffee-house and there, among others, had a good discussion with an Iron-merchant . . . Then I went and sat by Mr. Harrington and some East country merchants; and talking of the country about Quinsborough [Königsberg, E. Prussia] and thereabouts . . . Mr. Harrington told us how they do to get so much honey as they send abroad. They make hallow [sic] a great Firr tree, leaving only a small slitt down straight in one place; and this they close up again, only leave a little hole and there the Bees go in and fill the bodies of these trees as full of wax and honey as they can hold; and the inhabitants at their times go and open that slit and take what they please, without killing the bees, and so let them live there still and make more.

Diary entry for 5 May 1665

After dinner to Mr. Evelings: he being abroad we walked in his garden, and a lovely noble ground he hath endeed. And among other rarities, a hive of Bees; so, as being hived in glass, you may see the Bees making their honey and combs mighty pleasantly.

Two entries mention metheglin:

29th February 1659/60

So to my mother's and then to Mrs. Turner's . . . Here my Cosen Norton gave me a brave cup of Metheglin, the first I ever drank.

25th July 1666

[Mr. Williams] left me there among two or three of the King's servants, where we dined with the meat that came from his table; which was most excellent, with most brave drink, cooled in Ice (which at this hot time was welcome); and I, drinking no wine, had Metheglin for the King's own drinking, which did please me mightily.

There have been numerous editions of Pepys' *Diary*. The above extracts have been taken from *The diary of Samuel Pepys*, a new and complete transcription ed. Robert Latham and William Matthews, London, Bell, 9 vol. 1970–76, *Companion* and *Index* vol. 10, 11, to be published.

Evelyn was given a 'transparent apiary' by Dr. Wilkins, Warden of Wadham College, Oxford, on 13 July 1654. Charles II visited Evelyn's garden at Sayes Court on 30 April 1663 and admired the hive. Evelyn was at home on 5 October 1665 and showed Pepys his garden but there is no mention of the bee hive. A drawing of the hive is illustrated in

Elysium Britannicum (34, 774). These early transparent hives would only show the outside combs. The first known single-comb transparent hives were noted by Maraldi in the garden of the French Royal Observatory in 1687; *see* Fraser's *History* (635) p. 39.

POWER, Henry 38 1663

Experimental philosophy . . . containing new experiments microscopical, mercurial, magnetical./London, T. Rowcroft for J. Martyn & J. Allestry, 1663./The bee. Book 1 obs. 11 pp. 3–4. Book 2 obs. 11 and 17./A

Wing P3099.

Quoted in Blount's *Natural history* (66).

EVELYN, John 39 1664

Kalendarium hortense, or, The gardener's almanac, directing what he is to do monthly throughout the year./London, Macock for Martyn & Allestry, 1664./A O R. W

Many eds. to 10th./1706./A N O

Wing E3491–3502./Cole 465./Henrey 127–131./Hunt 296, 383, 411.

There are many references to bees and these, together with the references to bees in Evelyn's diary, are collected in the edition of *Elysium Britannicum* published in 1966 (774); *see* Evelyn (34).

HOOKE, Robert 40 1665

Micrographia, or, Some physiological descriptions of minute bodies made by magnifying glasses, with observations and enquiries thereupon./London, J. Martyn & J. Allestry, 1665./pp. 163–164. illus./A

Micrographia restaurata (or The copper plates of Dr. Hooke's . . . discoveries by the microscope reprinted and fully explained)./London, 1745./Sting of a bee Plate XVIII Fig. 2 pp. 30–31./A

Reprint./Weinheim, Cramer; also Codicote (Herts.), Wheldon & Wesley; also New York, Hafner, 1961, Historiae Naturalis Classica Vol. XX.

Wing H2620–21./Lisney 37–41./Cole 612–613.

Hooke's drawing of a bee's sting was the first study of an internal organ of the bee made with a microscope. See note on Cesi (1625) under Thorley (97). *See also* Leewenhoeck (47).

[BLAGRAVE, Joseph] 41 1669

Epitome of the art of husbandrie by J. B., Gent./London, for B. Billingsley & O. Blagrave, 1669./Section on bees./A O

2nd ed./London, for B. Billingsley, 1670./A R

3rd ed./London, 1675./Bees pp. 64–66./A N O R S U. W

4th ed./London, by B. Billingsley, 1685./A O P R S

Wing B3115–18.

Systema
Agriculturæ,
Being
the Mystery Of Husbandry
Discovered and
layd Open
by
J W
H Van Hove Sculp
Printed for Tho. Dring at y Corner. of Chancery lane in Fleetstreet. 1675.

Born at Reading, Blagrave was a mathematician of some note and a great enthusiast in astronomical studies; he also wrote much on agriculture. The bee matter is largely reproduced from Fitzherbert's 1523 book (1) without acknowledgement.

DIGBY, Sir Kenelm 42 1669

The closet of the eminently learned Sir Kenelm Digbie, Kt, opened: whereby is discovered several ways for making of metheglin, sider, cherry wine, etc. / London, E.C. for H. Brome, 1669. / Metheglin, meathe, hydromel pp. 1–126. / A O

2nd ed. / 1671.

3rd ed. / 1677. / A

New ed. *The closet of Sir Kenelm Digby, Knight, opened.* / London, Warner, 1910. / lv, 291 p., introduction, notes. glossary. / ed. by Anne Macdonell. / A R. C W

Wing D1427–29.

'Hydromel as I made it weak for the Queen Mother.'

B has photocopy of the 1910 ed. pages which refer to honey, mead, metheglin.

[WORLIDGE, John] 43 1669

Systema agriculturae, or, The mystery of husbandry discovered . . . to which is added *Kalendarium rusticum* [containing monthly hints to bee-

---

Fig. 4. Frontispiece of *Systema agriculturae* by John Worlidge (1669, entry 43). *The Explanation* commences:

First cast your Eye upon a Rustick Seat,
Built strong and plain, yet well contriv'd, and neat;
And scituated on a healthy Soyl,
Yielding much Wealth with little cost or toyl.
Near by it stand the Barns, *fram'd to contain*
Enriching stores of Hay, Pulse, Corn *and* Grain;
With Bartons *large, and places where to feed*
Your Oxen, Cows, Swine, Poultry, *with their breed.*
On th'other side, hard by the House, you see
The Apiary *for th' industrious* Bee.
Walk on a little farther, and behold
A pleasant Garden *from high Windes and Cold*
Defended (*by a spreading fruitful* Wall,
With Rows of Lime *and* Fir-tree *streight and tall,*)
Full fraught with necessary Flow'rs *and* Fruits,
And Natures choicest sorts of Plants *and* Roots.

The apiary, with 15 skeps, is to the right of the house.

keepers] and *Dictionarium rusticum* published for the common good by J. W., Gent. / London, printed by T. Johnson for S. Speed near the Inner Temple gate in Fleet Street, 1669. / 17 leaves, 278 p. illus. 29 cm. / Bees pp. 156–175. / A N O S. C

2nd ed. / London, for T. Dring, 1675. / Bees pp. 168–188. illus. / A G O S. W

3rd ed. / London, T. Dring, 1681. / Bees pp. 179–197. / A D G O R S. W

4th ed. / London, T. Dring, 1687. / Bees pp. 179–198. / A G R. C W

Anr. ed. The fourth edition carefully corrected and amended. / 1697. / A S

Anr. ed. *A complete system of husbandry and gardening*, or, The gentleman's companion in the business and pleasure of a country life. / London, Pickard, 1716. / Bees pp. 277–311. / A G O R S. M W

This is the last edition of *Systema agriculturae*, and the bee matter is little altered.

Wing W3598–3602.

The *Kalendarium rusticum* is itself dated 1668, but presumably did not appear until 1669. B holds a transcript of the references to bees.

The *Dictionarium rusticum* is a brief 12-page glossary and is quite different from 71. B also has *The second parts of Systema agriculturae . . . and Vinetum Britannicum*, published in 1689, but containing no bee matter. It is bound with *Vinetum Britannicum* (54).

Fig. 4 shows the frontispiece of the 1st edition, reproduced from the copy in the University of Reading Library.

CORNEY, Richard 44 1670

Some observations experimental touching bees. by Richard Corney of Harrington, Northamptonshire. An unpublished manuscript in the library of the University of Guelph, Canada. / 29 p.

B has a photocopy and transcription.

Richard Corney, Rector of Harrington, was with some difficulty induced to dictate his experiences and asked that they be presented to Mr. Thomas Isham, eldest son of Sir Justinian Isham. The 'much beloved reverend old person' died not long after. The manuscript gives a brief account of beekeeping with straw hives. Corney placed a second hive under the first so that he could take honey from the lower hive without driving the bees. He considered the drone to be the male bee 'by two little white stones (or testicles) he hath', but thought that the old queen stayed behind when a swarm left.

WOLLEY, Hannah 45 1670

The queen-like closet, or, Rich cabinet, stored with all manner of rare receipts for preserving, candeying and cookery. Very pleasant and

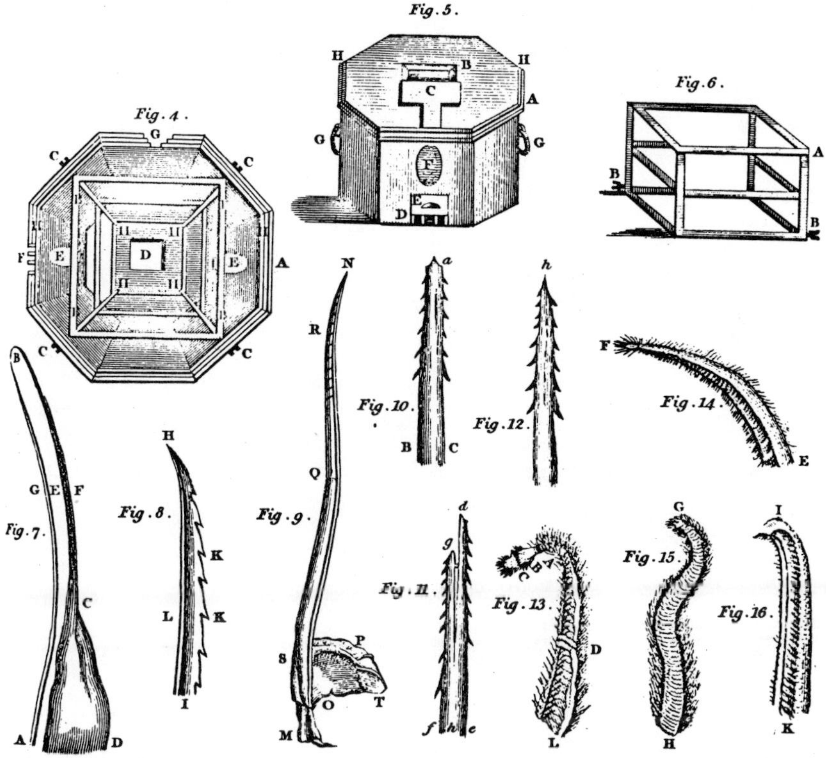

Fig. 5. Diagrams of a hive, and of parts of bees observed with the aid of a microscope, *Philosophical Transactions of the Royal Society of London*, 1673 (1809 reprint).

Figs. 4–6—A bee-house [octagonal hive-body] used in Scotland. The boxes (5) can be superimposed, the hole B with its shutter C regulating movement of bees between boxes. The framework (6) is placed inside the box (5), viewed from below in 4, 'for the bees to fasten their work upon'. The beekeeper cuts out the frame 'with all the bees' work upon it', by manipulating a double-bladed knife.

Figs. 7–12—Parts of the bee's sting.

Figs. 13–15—Several members about the head of a bee: one of the two arms (13), one of the two scrapers (14), and the wiper (15).

Fig. 16—The scraper of a wild bee.

Figs. 4–6 show Gedde's hive (Sir William Thomson, entry 47a).

Figs. 7–16 are drawings by A. van Leewenhoeck (entry 47).

beneficial to all the curious persons of the female sex./London, Lownes 1670./Metheglin p. 184./A

2nd ed./1672./A

3rd ed./1675./Metheglin p. 160. How to make wax work and colour wax pp. 182–184./A

4th ed./1681./A

5th ed./1684./A O

Wing W3282–86./Hunt 371.

BEAL, John 45a 1671

An apiary or Discipline of bees . . ./Philosophical Transactions of the Royal Society 6(71): 2148–2149 (1671)./A (B) P

B., F. 46 1672

The office of the good housewife, with necessary directions for the ordering of her family and dairy . . . also the manner of keeping and governing silk-worms and honey-bees: both very delightsome and profitable./ London, printed for Richard Mills at the Pestle and Mortar without Temple Bar, 1672./Bees pp. 97–143./A

2nd ed./1696./A

Wing B63.

LEEWENHOECK, Antony van 47 1673

A specimen of some observations made by a microscope, contrived by M. Leewenhoeck in Holland, lately communicated by Dr. Regnerus de Graaf./*Philosophical Transactions of the Royal Society* 8(94): 6037–6038, 8(97): 6116 (1673)./A (B) P

Descriptions and drawings of the sting and mouth parts. The second paper is entitled *The figures of some of Mr. Leewenhoeck's microscopical observations . . . together with their explication*. See Fig. 5.

THOMSON, Sir William 47a 1673

A description of a bee-house, useful for preventing the swarming of bees, used in Scotland with good success./*Philosophical Transactions of the Royal Society* 8(96): 6076–6078 (1673)./A (B) P

See Fig. 5, also note to next entry.

GEDDE, John 48 1675

A new discovery of an excellent method of bee-houses and colonies, to free the owners from the great charge and trouble that attends the swarming of bees, and delivers the bees from the evil reward of ruine for the benefit they brought their masters; advantaging their owners many-fold above what-ever any method heretofore practic'd doth. Experienced seven years by John Gedde, Gent., inventor; and approved by the Royal Society at Gresham Colledge./London, printed for the author and sold by D. Newman, 1675./30 p. 1 pl. 14 cm./A O P R S. M

2nd ed. enl. and several objections answered./London, Author, 1676./ 63 p. 1 pl. 14 cm./M

3rd ed. enlarged./London, Author, 1677./77 p. 1 pl. 14 cm./O. C M

3rd ed. [sic] *A further experimental discovery of bees and their usefulness.*

The third edition enlarged./London, Author, 1697./51 p. 2 pl.

Reprint of the 1st ed. is included in *The English apiary* (81).

Wing G443.

Gedde claimed to have invented his own particular bee hive. A patent was taken out in 1675, no. 180, in the names of Gedde, Galt and Nowell. We have consulted the patent specification in the Patent Office Library, but it gives very little detail about the hive itself. Gedde complains in his introduction that the Royal Society had printed an account of his patented bee boxes (*see* 47a). The Society had obtained the account from Sir Robert Murray; he had received it from Sir William Thomson, to whom Gedde had communicated it for his private information. Hartlib, *The reformed commonwealth of bees* (33), gives details of a similar hive designed by Rev. Mew, Vicar of Eastlington, Glos. Worlidge (51) thought that the patent stood in the way of progress. In 1679 the sale of Gedde's hives was taken over by Rusden (55).

An account of Gedde's life by R. Mylne appeared in *Miscellany of the Abbotsford Club* Vol. I, Edinburgh, 1837. See also an article by Walker in B.B.J. 50: 265–267, 306–307, 401–402, 438–439, which gives further bibliographical details (1922).

PLOT, Robert 49 [1676]

The natural history of Oxfordshire./Oxford, the Author printed at the theater in Oxford, and are to be had there, and in London at Mr. Moses Pits, and at Mr. S. Millers, [1676]./Bees pp. 24, 180–182.

Anr. ed./1677./A D N O R U

2nd ed./1705./A O P U

Facsimile reprint of 1705 ed./Chicheley (Bucks.), 1972.

Wing 2585–86./Cole 888.

Mentions Bishop Wilkins' hive. *See* Hartlib (33).

[WORLIDGE, John] 50 1676

Apiarium, or, A discourse of the government and ordering of bees, tending to . . . the discovery of the fallacies that are imposed by some, for private lucre, on the credulous . . . written by J.W., Gent./London, printed for and sold by Thos. Dring, 1676./28 p. 1 pl./A G O

2nd and later ed. with *Vinetum Britannicum* (*see* 54).

Anr. ed. *Compleat bee-master.*/London, Conyers at the Ring in Little Britain, 1698./28 p. 1 pl./A. M

Anr. ed. *New and compleat bee-master.*/[1720.]/12 p. 1 pl./A

Although there is no reference to Worlidge as the author, this is a revised version of *Apiarium* in smaller type with an almost identical woodcut. Presumably the publisher is Conyers, because there is a reference to Butler's book 'sold for 1s. at the Ring in Little-Britain'. The publisher is using the book to issue with other tracts. A's copy is bound with two other 12-page works, *Secrets of art and nature or modern curiosities* and *Directions how to keep horses and sheep free from all manner of diseases* by W.M. To this latter, two pages on bees have been added: 'A late Author affirms, that the Drone Bee is the Male, and that his Business is to sit upon and hatch the Eggs, and nourish the Young, whilst the Females follow their Work.'/

Wing W3594–96.

Walker p. 51: 'The New Book of Notable Things . . . III. The Compleat Bee-master. Lond[on], 1749. About 200 pp. of subjects separately paged; Bee Treatise, 12 pp. . . . founded on Worlidge's 'Complete Bee-master', with the same wood-cut.'

[WORLIDGE, John] 51 1677

Systema horticulturae or The Art of gardening by J.W., Gent./London, T. Burrel & W. Hensman, 1677./A

There were further editions in 1683 and 1688./A

Wing W3603–06A./Henrey 369, 374–376./Hunt 380.

In his Preface to the Reader Worlidge attacks Gedde, implying that his patent stood in the way of progress: 'Many ingenious men having undertaken at their great cost and pains these two last years, and are yet upon their experiments, to see the end of it, very much to the reputation and advantage of him that gives them leave so to do.' See Fraser's *History* p. 40 for a fuller quotation.

There were further editions in 1683 and 1688./A

LEMERY, Nicholas 52 1677

A course of chymistry, translated from the 5th edition of the French by W. Harris M.D./London, R.N. for W. Kettilby, 1677./D O

2nd ed. enl./London, R.N. for W. Kettilby, 1686./Of honey p.542 Distillation of wax pp. 545–548./A O

3rd ed./1698./D N

4th ed./1720./A O

Wing L1038–40.

SALMON, William 53 1678

Pharmacopoeia Londinensis, or, The new London dispensatory; translated into English for the publick good, and fitted to the whole art of healing./London, Dawks, 1678./Apes pp. 254–256./A O

Eds. in 1685./A./1691./O./1696./U./1702./O./1710./D.

Wing S437–441.

Honey is recommended as a resister of putrefaction. Salmon was the translator.

WORLIDGE, John 54 1678

Vinetum Britannicum, or, A treatise of cider and other wines . . . to which is added a discourse teaching the best way of improving bees. 2nd ed./London, pr. for Thos. Dring & Thos. Burrell, 1678./42 p. illus. 1 pl./A G H N R S. C M

The 'discourse' is *Apiarium* (*see* 50), which was not included in the 1st ed. (1676).

3rd ed./London, for T. Dring, 1691./11 leaves, 236 p. front; 4 leaves, 42 p. front./A B G P. M

Wing W3608–10./Hunt 348, 384.

There is a brief mention of metheglin or hydromel, pp. 168–173.

B's copy is bound with *The second parts of Systema Agriculturae* . . . and *Vinetum Britannicum*, published in 1689, which has no bee matter. *See* 43.

RUSDEN, Moses 55 1679

A further discovery of bees. Treating of the nature, government, generation and preservation of the bee. With the experiments and improvements arising from the keeping them in transparent boxes, instead of straw-hives. Also proper directions . . . as well to prevent their robbing in straw-hives, as their killing in colonies. Published by His Majesties especial command, and approved by the Royal Society at Gresham College./London, printed for the Author and are to be sold at his house next the Sign of the King's Arms in Bowling-Alley near the Abby in Westminster: and by Henry Million, bookseller, at the Bible in the Old-Bayley, 1679./12 leaves, 143 p. illus./A B G H O P R S U. C M W

2nd ed. *A full discovery of bees* . . . /London, H. Million, 1685./illus./A G. C

Anr. ed./London, pr. for Nathaniel Ranew at the King's Arms in St. Pauls Churchyard, 1687./. C

Wing R2312–13.

Rusden was an apothecary, and a member of the Apothecaries' Society in London. The original entries of his apprenticeship and of his being examined and admitted a member of the society are in the Guildhall Library, London. He was appointed 'Beemaster to the King's most excellent Majesty' at the instigation of John Evelyn. Rusden's claim to this title had been doubted, and Houghton (69) said 'He stiles himself Bee master to the King's excellent majesty viz Charles II'. However, Evelyn's manuscript (34) vindicates Rusden's claim. The transparent boxes are the same as Gedde's transparent octagon. Rusden's plate of his hive and the Royal coat of arms is reproduced facing p. 81 in Fussell, *The old English farming books, 1523–1730*.

MATHER, William 56 1681

The young man's companion, or, A very useful manual for youth. Plain and easy directions for spelling, reading and writing English . . . with easy rules for arithmetick and the Englishing of the Latin Bible without a tutor. / London, T. Snowden, 1681. / The ordering of bees pp. 276–283. / A

The book went through many editions. We have notes of the following:
8th ed. / 1710. / Bees pp. 203–208. Mead p. 192. / A
15th ed. / 1737. / Mead p. 328, nothing on bees. / A

Wing 1287, 1287A.
Mather was a schoolmaster of Bedford.

WHELER, Sir George 57 1682

A journey into Greece. / London, for T. Cademan, 1682. / Greek hive pp. 411–413. illus. / A D N O S U. M

Wing W1607.

Account of beekeeping on Mount Hymettus and illustration of Greek skeps with movable bars. This is probably the first account of a movable-comb hive published in Europe. Wildman's skeps with bars (119) were similar, but Wildman makes no mention of the advantages of being able to remove individual combs. Wheler was born in Breda, Holland. He became Prebendary of Durham, and there is a memorial to him in Durham Cathedral.

A., J. 58 1683

Directions for the making of colonies for bees . . . from *A collection of letters for the improvement of husbandry and trade* edited by J. Houghton. / London, sold by the Author, Lawrence and others. / In 2 vol. / The work was published periodically 'in small parcels often' 1681–83. / A (B) O S. M W

The bee matter is contained in pp. 166–174 of Vol. 2 No. 6, dated 16 June 1683.

Anr. ed. rev. and corr. by R. Bradley./London, Woodman, 1728./Bees pp. 374–381./A

2nd ed. rev. and corr. by R. Bradley./London, Woodman, 1728./A

Except for the title page these two editions appear identical. They were issued under the title *Husbandry and trade improv'd* (see below).

The *Collection of letters* . . . should not be confused with *A collection for improvement of husbandry and trade* (69) also edited by Houghton and issued periodically 1692–1703. In 1727 Bradley revised and corrected the *Collection for improvement of husbandry and trade* omitting the weekly market reports and advertisements. This was reissued in the same year under a new title *Husbandry and trade improv'd*, and a 4th volume added in 1728 which was in fact a new edition of the *Collection of letters.*

Directions for the making of colonies for bees was reprinted in *B.B.J* 56: 453–456 (1928) by Walker, and this article was reprinted in 1930 in Herrod-Hempsall's *Beekeeping new and old* Vol. 1 (658), pp. 143–154.

As Walker says, this was the first well thought-out and publicly described design of a wooden hive with movable frames having top, side and bottom bars (but no bee space): It is tempting to think that John Aubrey may have been the author; he described the hive as 'a good modelle of a bee-hive' in his *Natural history of Wiltshire* (63) Chap. 13, p. 67.

REID, John 59 1683

The Scots gard'ner, in two parts . . . whereunto is annexed the gard'ners kalendar./Edinburgh, David Lindsay, 1683./6 leaves, 125 p. 1 leaf, 14 p. 4 pl./A (B) N R S. W (B's copy is a transcript of the references to bees in *The gard'ners kalendar.*)

2nd ed. rev. and enl./Edinburgh, Moncur, 1721./5 leaves, 124 p. 1 leaf, 54 p. 24 p. 4 pl. front./A. W

One of the additions is *The florists vade mecum.*

3rd ed. *The Scot's gardiner* for the climate of Scotland in 3 parts . . . together with the gardiner's kalendar, the florist's vade-mecum, the practical bee-master . . . now carefully corrected, with alterations and additions by an eminent hand./Edinburgh, Reid, 1756./xii, 208 p. vii, 40 p. viii, 138 p. 48 p. 4 pl./O S. W

Anr. ed./Edinburgh, Reid, 1766./viii, 208 p. 48 p. viii, 40 p. viii, 138 p. 4 pl./A R. W (A's copy lacks *The practical bee-master.*)

Anr. ed. ed. by A. H. Hyatt./1903./A

Anr. ed./1907./A O S/Garden Lover's Series

Wing R763–764./Henrey 328–331./Hunt 368, 591.

The practical bee-master included in the 1756 and 1766 eds. is the 2nd ed. 1750 of Maxwell's book (*see* 102). It has separate title page and pagination viii, 138 p.

[SMITH, John] 60 1684

Profit and pleasure united, or, The husbandman's magazine; being a most exact treatise . . . with directions for the encreasing and preserving of bees . . ./London, printed for J. Blare at the Looking-Glass on London Bridge, 1684./A O

Anr. ed. signed John Smith, Gent./1705./Of ordering and improving stocks of bees pp. 125–131./R

Wing S4095.

GARDEN, George 60a 1685

[*Letter concerning the proboscis of bees.*]/*Philosophical Transactions of the Royal Society* 15(176): 1156–1158 (1685)./A (B) P

George Garden was a doctor in Aberdeen.

PLOT, Robert 61 1686

The natural history of Staffordshire./Oxford, at the Theater, 1686./Osier hives pp. 385–387./A D O P R U. M
Wing P2588.

W., R. 62 1688

A necessary family book, both for city and country./London, Harris, 1688./How to keep cherries, pears, nuts or other fruit a whole year, as fresh as they came from the tree p. 72./A

Anr. ed./London, Rhodes, 1710./Directions for profitable management of bees pp. 62–65.

Wing W100.

AUBREY, John, F.R.S. 63 [c. 1690]

Natural history of Wiltshire . . ./Extracts; for Wiltshire Topographical Society. Edited by Britton./London, J. B. Nichols, 1847./Bees pp. 67–68./ A O

Reprinted with introduction by K. G. Ponting./Newton Abbot (Devon), David and Charles, 1969.

Many interesting references to Hampshire as the best county for honey, various hives including J.A.'s (see 58) and to a scale hive at Kilmanton. Refers to Dodoens and Lyte (see 10). Ref. to Charles Butler (see 18), who gave some bees to his daughter, the product of which, when she came to be married, was a £400 portion. She was an ancestress of Gilbert White of Selbourne. Evelyn saw the manuscript and added some notes. Aubrey started the book in 1656, and in 1690 he deposited a fair copy with the Royal Society; he never published it.

TRYON, Thomas 64 [1690]

The good housewife made a doctor, or, Health's choice and sure friend, being a plain way of nature's own prescribing, to prevent and cure most diseases incident to men, women and children, by diet and kitchen physick only. / London, [1690]. / O

2nd ed. / London, H.N. & T.S., 1692. / A O

Wing T3180–81.

Of honey, its nature and operation pp. 118–124. '. . . it is extracted out of the herbs and flowers into a little bag or thin skin which lies in the very center of the bee.

Y-WORTH, William 65 1690

The Britannian magazine, or, A new art of making above twenty sorts of English wines. / 1690. / A

2nd ed. by W.Y., M.D. / London, Onley, 1694. / Mead p. 41. / A
3rd ed. / for Bodington, [1700]. / A
Wing Y213–215.

BLOUNT, Sir Thomas Pope 66 1693

A natural history containing many not common observations, extracted out of the best modern writers. / London, for R. Bentley, 1693. / Bees pp. 293–303. / A D G O
Wing B3351.

HARTMAN, George 67 1696

The family physitian . . . containing some hundred of considerable receipts and secrets of great value, with observations of great cures, together with the true English wine-cellars: or The right method of making English wines, or metheglin . . . by Geo. Hartman, Phylo Chymist, who liv'd and travell'd with the Honerable and Renown'd Sir Kenelm Digby in several parts of Europe, the space of seven years till he died. / London, Wellington, 1696. / Meads, metheglin & hydromel pp. 492–497. / A
Wing H1003.

VIRGIL (Publius Vergilius Maro) 68 1697

The works of Virgil containing his pastorals, Georgics and Æneis. Translated into English verse; by John Dryden. / London, for Jacob Tonson, 1697. / 24 leaves, 640 p. 101 pl. front. / Fourth *Georgic* Vol. 1 pp. 122–147, 5 pl. / A B D O

Subsequent eds. 1701, 1709, 1716, 1721, 1730. / A G S. M

Wing V616.

Fig. 6. Engraving from a design of Wenceslaus Hollar, in John Dryden's 1697 translation *The works of Virgil* (entry 68). The lines referring to the illustration are:

But when thou seest a swarming Cloud arise,
That sweeps aloft, and darkens all the Skies:
The Motions of their hasty Flight attend;
And know to Floods, or Woods, their airy march they bend.
Then Melfoil beat, and Honey-suckles pound,
With these alluring Savours strew the Ground;
And mix with tinkling Brass, the Cymbals droning Sound.
Streight to their ancient Cells, recall'd from Air,
The reconcil'd Deserters will repair.

The book is noted for its engravings (Fig. 6). Two of the three bee illustrations were engraved from designs of Wenceslaus Hollar who died in 1677; they may have been used in earlier editions, in the original Latin. Below one is a dedication to the 'subscriber to the cut', the Right Honble. S^{ir}. William Trumbull, K^{t}.; his coat of arms shows, in two of the quarters, a skep of the same shape as those in the engraving.

B's copy was a presentation by Mrs. M. B. Townley on the occasion of the 25th anniversary of the Bee Research Association.

1700 to 1799

Scientific observation and practical experience

HOUGHTON, John, F.R.S. *editor* 69 1702

A collection for improvement of husbandry and trade./A periodical published in 583 weekly issues divided into 19 volumes, 1692–1703./38 cm./ Three issues published in 1702 contain material on bees: 524 *Bees describ'd. Who has writ of them. Box hives. Straw hives. Sorts of bees. Oeconomy art and industry.* 525 *Farther about bees.* 526 *How much honey, wax etc. imported Anno 1694.*/A

Anr. ed. rev. and corr. by R. Bradley./London, Woodman, 1727./3 vol. Bees Vol. 3 pp. 257–262./A G O

Anr. ed. under the title *Husbandry and trade improv'd*, rev. and corr. by R. Bradley./London, Woodman, 1727./3 vol.

Husbandry and trade improv'd, 2nd ed.. rev. and corr. by R. Bradley./ London, Woodman, 1728./3 vol./A

Bradley's three editions appear to be identical except for the title page. For the 4th volume see under 58.
The entire series was republished by Gregg International Publishers, Farnborough, Hants., in 1969.

MOORE, Sir Jonas 70 1703

England's interest, or, The gentleman and farmer's friend, shewing . . . of the husbandry of bees, and the great benefit thereby. 2nd ed. with large additions./London, J. How, 1703./Bees pp. 95–161./14 cm./A O R

3rd ed. with large adds./1705./A G N. M
4th ed./1707./G. C
Anr. ed. with adds./1721./A G P. M W

The first edition appears not to have survived; Richard Weston's book, in 1769 (122), states that it was published in 1695. The entire chapter on bees was reprinted in Cotton's *My bee book* (238) from the 1707 edition, and Walker states that the various editions are similar. Moore (1617–1679) was a distinguished civil servant and was appointed surveyor to carry out the draining of the Fens. This book was a posthumous work.

71 1704

Dictionarium rusticum et urbanicum; or, A dictionary of all sorts of country affairs, handicraft, trading, and merchandising . . . the breeding, feeding and managing of all sorts of cattle, as also of bees . . . /London, for J. Nicholson, 1704./approx. 800 unnumbered pages, illus./A O R S U. W

2nd ed./*Dictionarium rusticum, urbanicum et botanicum* rev. corr. & improved./London, for J. Nicholson & others, 1717./A G R S. M

3rd ed. rev. corr. & improved./London, for J. & J. Knapton & others, 1726./A G O R. C

Hunt 439.

As the title implies, the book is a dictionary; it has entries such as apiary 3 pp., bee-hives 3 pp., bees 6 pp., bees stinging 1 p., driving of bees 2 pp., exsection or gelding of combs 1 p., generation of bees (from a calf or an ox) 1 p.

The bee matter is taken chiefly from Worlidge, but also from Hartlib, Virgil, etc. Some list this work under Worlidge, and it is often catalogued under Nathaniel Bailey, who compiled the *Dictionarium domesticum* (90).

[MORTIMER, John, F.R.S.] 72 1707

The whole art of husbandry . . . to which is added the country-man's kalendar, what he is to do every month in the year./London, J.H. for H. Mortlock, 1707./Bees pp. 200–218, mead pp. 605–608./A O R U. M W

2nd ed./London, Mortlock & Robinson, 1708./Bees pp. 201–218. /A N R. C

3rd ed./1712./Bees pp. 191–208./R S

4th ed./1716./Bees pp. 266–288./O R S. W

5th ed./Dublin, Grerson, 1721./Bees pp. 168–181./A G R. W

6th ed. rev. by T. Mortimer./1761./A R. C W

Fussell, *The old English farming books*, p. 96: 'John Mortimer, F.R.S., was a merchant in London on Tower Hill, until he came into an estate in Essex called Topping Hall in 1693. He was an enthusiastic agriculturist. . . .'

73 [1710]

The compleat English and French vermin-killer: being a companion for all families. shewing a ready way to destroy adders, badgers . . . wasps . . . with some directions for gardiners, and the prizes of workmens labour, being a rich cabinet of modern curiosities./London, Conyers, [1710]./ 70 p. illus. 14 cm./A

Directions are given for destroying wasps and hornets to prevent them from damaging bees and fruit.

Copy in A contains a catalogue of books sold at the Ring in Little Britain by G. Conyers. They include:

The way to get wealth . . . to make mead, rum, rack, mum, coffee, tea, chocolet, and 40 sorts of ale in a minute . . . 1s. 6d. It states that the following are often sold with it:

The way to save wealth: to live for two pence a day . . . to destroy all sorts of vermine; to order silk-worms and bees . . . 1s. 6d.

A thousand notable things . . . to catch birds, angle, kill vermine . . . preserve bees . . . 1s. 6d.

We have not seen these books but suspect that the bee matter is based on Worlidge's *Apiarium* (50).

Anr. ed. *The vermin-killer.*/London, Bristow & others, [1750]./72 p. no illus. 17 cm./A

This ed. contains the article about bees referred to in Fraser's *History*, which quotes (p. 64) the passage dealing with hunting bees' nests. 'And to know whether they are far or nigh, set sugared Water and, when they come to sip, sprinkle some Red, Green or other Colour upon them, and by their often or seldom coming, you may guess the Distance or Nearness of their Nests; as also by the sudden Resort of great Numbers, so soon as one or two have found out the Sweets, for they suddenly give notice to their Fellows.'

Anr. ed. *The compleat vermin-killer* 2nd ed./London, Fielding & Walker, 1777./88 p. no illus. 22 cm. Bee article omitted./A

Anr. ed. *The new and complete vermin-killer.*/London, Dean & Munday, [1826]./iv, 60 p. illus.

'To preserve bees, fruits &c. from the attack of wasps and hornets.'

WARDER, Joseph 74 1712

The true amazons: or, The monarchy of bees: being a new discovery and improvement of those wonderful creatures. Wherein is experimentally demonstrated, I. That they are all governed by a queen. II. The amazing beauty and dignity of her person. III. Her extraordinary authority and power. IV. Their exceeding loyalty and unparallelled love to their queen. V Their sex, male and female. VI. The manner of their breeding. VII. Their wars. VIII. Their enemies, with directions plain and easy how to manage them, both in straw-hives and transparent boxes; so that with laying out but four or five pounds, in three or four years, if the summers are kind, you may get thirty or forty pounds *per annum.* Also how to make the English wine or mead, equal, if not superior to the best of other wines. By Joseph Warder of Croydon, Physician./London, printed by I. Dawks for, and sold by John Pemberton, at the Buck and Sun over against St. Dunstans Church in Fleet-Street, 1712./xii (p. xi wrongly numbered ix) 2 leaves, 166 p./A B G S. C M W

2nd ed. with adds./London, Pemberton, 1713./This and later eds. have a frontispiece with a portrait of the author./xii, 2 leaves, 160 p. front. /A B G H N R S. M

3rd ed. with adds./London Pemberton, 1716./xiii, 1 leaf, 120 p. front. A B G O P. C W

4th ed. corr./London, Pemberton, 1720./xiii, 1 leaf, 120 p./A G O S U. M

This ed. appears not to have the frontispiece. See note at end of entry.

5th ed. to which is added a letter from the author concerning a late treatise upon the subject of bees./London, Pemberton, 1722./xiii, 1 leaf, 120 p. front./A G R S

The letter was in answer to the charge made by Nourse in *The English apiary* (81) that Warder had 'largely transcribed from others'. Warder rejects the charge indignantly, but in fact his hive is somewhat similar to Gedde's 'transparent octagon'.

6th ed./London, Pemberton, 1726./Contents as in 5th ed. but paged xxiv, 112 p. front./A B G S

7th ed./Letter from author is discarded./London, Longman, 1742./ 164 p. front./A G. W

8th ed./London, Longman, 1749./164 p. front./A G S. C

9th ed./London, Baldwin, 1765./164 p. front./A B G S U. C M W

An important and successful book, which went through many editions and was translated into German and Italian. Warder was permitted to dedicate the second edition to Queen Anne. Fraser told us that, according to the Secretary of the College of Physicians, Warder was qualified as an Extra Licentiate of the College of Physicians but that he obtained his qualification two years after he set up his plate in Croydon. Extra Licentiates could practise anywhere except in London. Walker p. 74 refers to an old manuscript note in his copy of the 5th edition: 'The author of this tract has been a Presbyterian, a Soldier, a Quack, and everything. But for Bees, I have been acquainted by some Gentlemen who have had ye curiosity to visit his Bee-hives, that no man in ye world did probably understand them better.'

Walker's copy of the 4th ed. (now in M) does have the frontispiece, but we have seen 4 copies and heard of 5 others, all without the frontispiece.

DERHAM, William 75 1713

Physico-theology, or, A demonstration of the being and attributes of God from his works of creation . . . being the substance of XVI sermons, preached . . . at Mr. Boyle's lectures, in the year 1711 and 1712 . . . with large notes and many curious observations never before published. /London, Innys, 1713./Picture of the sting and scraper (tongue)./A O

Many eds. to 18th in 1798.

Lisney 96–116./Cole 1252.

Derham (1657–1735) rivalled Butler in versatility. He wrote a similar demonstration from astronomy, also a book on clock and watch making. While he was Vicar, he did all the doctoring required in his parish (Fraser).

[MANDEVILLE, Bernard de] 76 1714

The fable of the bees, or, Private vices, publick benefits./London, Roberts, 1714./228 p./A D. M

Many eds. in 18th and 19th centuries./B has 6th ed. 1729. Later eds. include the following:

Anr. ed./ed. with commentary by E. B. Kaye./Oxford, Clarendon Press, 1924./A

Anr. ed./ed. with introduction by D. Garman./London, Wishart, 1934./A

Anr. ed./ed. with introduction by Phillip Harth./London, Penguin Books, 1970./A B

A political satire. The book sprang from *The grumbling hive, or, Knaves turn'd honest, a satire in verse*. London c. 1700. Issued as a sixpenny pamphlet, it contained four pages of doggerel describing a hive in which all the bees suddenly turn strictly honest and come to grief in consequence. In the book the idea is enlarged and applied to human society. The book caused much controversy because, as its subtitle implies, Mandeville suggests that private vices such as luxury bring public benefits by promoting trade. Mandeville's attack on charity schools, added to the 1723 ed., increased the book's unpopularity. In 1729 Mandeville continued the subject in *The fable of the bees, Part II*. Pirated edition of the pamphlet is in A. Mandeville studied physics in Holland and became an M.D.

STEVENSON, Rev. Henry [P.] 77 1716

The young gardener's director . . . to which is prefixed some directions for the management of bees, to those that think fit, or have already introduced them into the garden./London, Baker, 1716./A

2nd ed. *The gentleman gardner's director* . . ./London, Austen, 1744./Bees pp. 267–273./G

3rd ed./London, Austen, 1748./Bees pp. 307–312./G. W

4th ed./1750./Bees pp. 307–312./G

5th ed. including *Oeconomical observations on bees* by M. du Hamel of the Royal Ac. Sciences, Paris./London, Hinton, 1764./A S U. M

6th ed. not seen.

7th ed./London, Hinton, 1766./A P S

8th ed. *The gentleman gardener instructed* in sowing, planting . . . and the management of bees . . . To the foregoing management of bees we shall beg leave to add, for the further instruction of the reader, the following curious and interesting Oeconomical observations on bees by M. du Hamel, a member of the Royal Academy of Sciences at Paris./London, J. Hinton, 1769./Mead pp. 274–275. Bees pp. 276–293./A

Henrey 1387–92./Hodgson 448./Hunt 584.

The first ed. has only the initials H.S. on the title page, but the preface is signed H. Stevenson, and the author's name appears on the title page of the later eds. Rev. Henry P. Stevenson was Master of the Free School of East Retford, Notts.

BRADLEY, Richard, F.R.S. 78 1718
The country gentleman and farmer's monthly director . . . with several particulars relating to the improvement of bees never before made publick. /London, Woodman & Lyon, 1718./Bees various pages./G

2nd ed./London, Woodman & Lyon, 1726./Bees pp. 17–18, 43–44, 96–100, etc./A D O R U. W
3rd ed./London, Woodman & Lyon, 1727./Mead p. 136./A O U. M
6th ed. enl./London, Browne, 1736./Bees pp. 20, 58, 85, etc./A D G. W

Extract (taking off a honey box): 'The man who performs this operation will save himself best from the bees if he has no hair on his head, and that the head be rubbed with vinegar.'

Fussell pp. 106–107 quotes the following thumbnail sketch of Bradley by Mitchell, *Wet days at Edgewood*, 1883: 'Richard Bradley was the Dr. Lardner of his day—a man of general scientific attainment, an indefatigable worker, venturing hazardous predictions, writing some fifteen or twenty volumes upon subjects connected with agriculture, foisting himself into the chair of Botany at Cambridge by noisy reclamation, selling his name to the booksellers for attachment to other people's wares, and, finally, only escaping the indignity of a removal from his professor's chair by sudden death in 1732.'

VIRGIL (Publius Vergilius Maro) 79 [1718]
Works of Virgil: translated by Richard Earl of Lauderdale into English verse. [By R. Maitland.]/London, Lintott, [1718]./Bees pp. 57–76./ A D O
A gives date [1709] and 2nd ed. [1717]

Richard Maitland was Earl of Lauderdale.

BRADLEY, Richard, F.R.S. 80 [1721]
A general treatise of husbandry and gardening . . ./London, Peele, issued in monthly parts April [1721]–March [1722].

August 1721 has 14 pp. on bees, including an account 'taken from some

Memoirs of the Royal Academy of Paris'. This seems to be based upon Maraldi who is mentioned by name.

September 1721 has 46 pp. on bees: 'An account of bees . . . with variety of curious observations relating to their oeconomy, by Mr. Maraldi . . .'/illus. This seems to be an expansion of the earlier account.

The series was continued under a different title for a further year.

The parts bound in 2 vol. with new title page./1721–22./A D G O S

Later copies with new title page./London, Woodward and Peele. 2 vol. Dated 1724 or 1726./Bees are in Vol. 1/A G O S U. M W

Anr. ed. *A general treatise of agriculture* . . . displaying the arts of husbandry and gardening . . . corrected and properly methodised . . ./London, W. Johnson, Baldwin & 12 others, 1757./A S

Henrey 485–87./Hunt 459.

Fraser's *History* p. 46 states: 'In 1742 an abridged English translation [of Maraldi's *Observations sur les abeilles* 1712] was published by John Martyn and Ephraim Chambers [*Phil. Hist. Mem. R. Acad. Sci.* 4: 168–177], but unfortunately no complete English translation of this valuable paper has ever appeared, although it contains the first accounts of many features of bee life which we now take for granted.'

DUDLEY, Paul 80a 1721

An account of a method lately found out in New-England, for discovering where the bees hive in the woods, in order to get their honey./*Philosophical Transactions of the Royal Society* 31(367): 148–150 (1721)./A (B) P

Abridged version./*Philosophical Transactions of the Royal Society* (*Abridged*) 7: 403–404 (1809)./A (B) P

GEDDE, John 81 1721–22

The English apiary, or, The compleat bee-master. Unfolding the whole art and mystery of the management of bees, being a collection and improvement of what has been written . . . relating to this subject . . . with a new discovery of an excellent method for making bee-houses and colonies . . . by John Gedde Esq./London, E. Curll, W. Mears and T. Corbet, 1721./108 p. front./A G H P R S. C M W

Walker p. 22: 'Front. pl[ate] (adapted from Gedde's); Title; then pp. v to ix, "To the Booksellers", a letter signed Cha. Nourse, in which Warder is accused of having "largely transcribed from others" [see Warder (74)]. Then: The Approbation of the Royal Society (not in Gedde's Treatise, and only the first Par[agraph] genuine); Gedde's "Introduction"; the body of the work (largely borrowed from Purchas); and finally, a correct reprint of the rest of Gedde's 1st. Edn., except that date of title is "printed in 1675 and reprinted in 1722".' There is a second title page following p. 82 which, as Walker says, is the same as the original 1675 title page except for the new date.

CHOMEL, Noel 82 1725

Dictionnaire oeconnomique, or, The family dictionary . . . containing . . . the most advantageous ways of breeding, feeding and ordering all sorts of domestick animals, as . . . bees etc., done into English, rev. and recommended by R. Bradley. / London, D. Midwinter, 1725. / 2 vol. / Articles on bees and honey. / A D G R S. M

Articles on apiary, bees, honey etc. Translated from *Dictionnaire oeconomique*, Paris 1718.

LAURENCE, Rev. John 83 1726

A new system of agriculture, being a complete body of husbandry and gardening in all parts of them. / London, Woodward, 1726. / xxiv, 456 p. illus. / Bees pp. 154–161. / A G N O R S. C M

Anr. ed. / Dublin, Hyde, 1727. / R. W

Henrey 945–946. / Hunt 466.

Laurence mentions an octagonal hive with glass panels and says that he is a kinsman of the inventor, Mr. Mew. However he is not very enthusiastic about the hive. 'I could never yet see the great Advantage boasted of from these *Cases* or Glass-Hives, either as to a Discovery of their Government, or the Increase of their Riches.' He says that it is difficult to see much through the glass, and that any system which depends upon taking away the bottom box after the bees have worked up to the box above, means that the honey thus obtained is in old rather than new comb. Laurence was Rector of Bishop Wearmouth, Co. Durham.

BRADLEY, Richard, F.R.S. 84 1727

The country housewife and lady's director in the management of a house and the delights and profits of a farm . . . 2nd ed. / London, Woodman, 1727. / To make mead and hydromel pp. 136–137. / A

1st ed. not seen. Later eds. 1728, 1732, 1736, 1762. / A O. M

[BOREMAN, Thomas] 85 1730

Description of above 300 animals . . . / London, Ware & Boreman, 1730. /

Bees pp. 204–207. illus. / A

37 other eds. to 1889, including Glasgow 1794, Edinburgh 1797, Perth 1807. / S

Lisney 126–166. / Cole 1322–1322c.

Contains a picture of bees and a box hive with straw top, and an account said to be extracted from Warder (74).

ELLIS, William 86 1731

Modern husbandman or, Practice of farming. / London, 1731. / Later eds. in 1742, 1743, 1750. / A G N O R. W

Scattered information on bees, and an account of moving bees on to the crop: 'At Cheddington, Bucks., a farmer carried out his swarms as soon as hived, and set them on stools among the field beans that were then in blossom, though the crop stood not thirty poles distant from the house. Here they got a great deal of honey in a little time, because they had not far to come and go . . .'

DUBLIN SOCIETY, by order of the 87 1733

Instructions for managing bees. / Dublin, Rhames, 1733. / 47 p. illus. / A (B) G O R S U. C M W

This first Irish bee book was a compilation intended to help farmers and others. It disapproved of Warder's boxes, and advocated straw hives and the use of a bee house in the damp Irish climate.

B's copy is typescript with traced illustrations.

PLUCHE, Noel Antoine 88 1733

Spectacle de la nature: or, Nature display'd. Being discourses on such particulars of natural history as were thought most proper to excite the curiosity and form the minds of youth. Translated from the French of 1732 by Samuel Humphreys. / London, J. Pemberton & others, 1733. / In 1 vol. in 2 parts xvi, 279 p., 300 p. with one index. / The book consists of a number of dialogues, two of which (Nos. 6 & 7 consisting of 38 p. with 1 pl.) are devoted to bees. / A O S. C W

2nd ed. / London, J. Pemberton & others, 1735. / 1 vol. in 2 parts xvi, 192 p., 207 p. with one index. / S. M

Fig. 7. Frontispiece of Nathaniel Bailey's *Dictionarium domesticum* (1736, entry 90), showing the mistress of a farm 'furnishing the house with the most common necessaries of life': milk and drink at the top, and fish and flesh at the bottom. It is a pity that the central picture does not show her in action also with the bees, instead of contemplating her herbs. The hives seen through the doorway are doubtless 'made of good wheat straw bound with bramble', as recommended in the text.

Frontispiece
Toms sculpt.

Between 1735 and 1739 a further 3 vols. were added. Vol. 3 contains a reference to bees as domestic animals, in the dialogue dealing with pasture and meadow grounds.
Subsequent eds./1736, 1739, 1740, 1744, 1754, 1757, 1776./A B (1776 ed.) S

Also published as *Nature delineated* translated by D. Bellamy./London, Hodges, 1739./G O S
Subsequent eds./1744, 1803./A S. M

Cole 1390./Hodgson 447./Hunt 499.

HARRISON, Mrs. Sarah 89 1733
The housekeeper's pocket book. London, T. Worrall, 1733./xii, 217 p. illus./A

2nd ed./London, R. Ware at Bible and Sun, Amen Corner, 1739./Meads pp. 193–194./A
Subsequent eds. to 9th./1777./A
Mrs. Harrison describes herself as being 'of Devonshire'.

BAILEY, Nathaniel 90 1736
Dictionarium domesticum, being a new and compleat household dictionary for the use both of city and country shewing . . . v. The apiary: or, The manner of breeding, hiving and managing of bees./London, Hitch, Davies & Austen, 1736./approx. 600 unnumbered pages with engraved front./A B O S

Some of the entries in this dictionary are clearly based on the *Dictionarium rusticum* (71). The entry on generation of bees is, however, omitted; possibly such a subject was considered unsuitable for the lady of the house. She is portrayed in the frontispiece (Fig. 7).

91 1736
The complete family-piece and country gentleman and farmer's best guide. Faithfully collected by several very eminent and ingenious gentlemen./London, Longman, 1736./Of bees and how to order them pp. 513–518. Mead p. 263./A

2nd ed./London, 1737./S

GAY, John 91a 1738
Fables. By the late Mr. Gay. Volume the second./London, Knapton, 1738./6 leaves, 155 p. illus./A

The *Fables* have been produced in over 350 editions, including recently a facsimile reprint of the 1727 and 1738 eds., with both vols. in one. London, Scolar Press, 1969 reprinted 1973/(B)

Also *John Gay, poetry and prose* ed. Winton A. Dearing. Oxford, 1974. 2 vols.

Gay, the author of *The beggar's opera*, published his first set of fables in 1727. The second set was published in 1738, six years after Gay's death, with plates drawn by Hubert François Gravelot and engraved by Louis Gérard Scotin; one of them shows a skep. The satire is more personal here than in the earlier fables. Fable X, *The degenerate bees*, is addressed 'to the Reverend Dr. Swift, Dean of St. Patricks' and, according to Armens' *John Gay, social critic*, New York 1954, the bee referred to in line 37 who sowed corruption throughout the hive was Horace Walpole. Dearing states that the fable may have been suggested by a fable of Phaedrus (III 13) in which the worker bees and the drones bring their differences of opinion to a wasp for judgment, or by de Mandeville (76). Certainly Gay's line 'By petty rogues the great ones thrive' is reminiscent of de Mandeville.

The British Library has one of the 525 copies of *The fox at the point of death* (one of the first set of fables) that was reprinted in secret on the private press of the Busy Bee in German-occupied Netherlands in March 1945.

DINSDALE, Joshua 92 1740

The modern art of breeding bees: a poem./London, Davidson, 1740./ 26 p./A G O S. M

Inspired by Virgil's Fourth *Georgic*.

Fraser's *History* p. 52 quotes 12 lines and says that Dinsdale 'managed to write rather better poetry than his title leads us to expect'.

VIRGIL (Publius Vergilius Maro) 93 1741

The Georgics of Virgil—with an English [prose] translation and notes by John Martyn./London, R. Kerby. 1741./

Vol. 1 xxii, 403 p. 29 cm./Bees pp. 321–403./A O. M

Eds. in 1746, 1755, 1811, 1827./A

Hunt 517.

Martyn was a Fellow of the Royal Society, and Professor of Botany at Cambridge in succession to Bradley (*see* 78). He produced a similar translation of the *Eclogues* in 1749.

BAKER, Henry 94 1742

The microscope made easy./London, R. Dodsley, 1742./Bee sting p. 209. Pl. XIII./A

Subsequent eds./1743, 1744, 1754, 1769.

Cole 1456–57. 1461.

95 1742

The curious traveller: being a choice collection of very remarkable histories, voyages, travels, etc . . . digested into familiar letters and conversations . . . containing among many other particulars . . . a curious account of the commonwealth oeconomy and natural qualifications of bees./London, Rowland, 1742./Bees pp. 219–234./A S

MacLAURIN, Colin 95a 1743

Of the bases of the cells wherein the bees deposite their honey./*Philosophical Transactions of the Royal Society* 42 (471): 565–571 (1743)./A (B) P

Abridged ed./*Philosophical Transactions of the Royal Society* (*Abridged*) 9: 2 only (1809)./A P

[BAZIN, Gilles Augustin] 96 1744

The natural history of bees. Containing an account of their production, their oeconomy, the manner of their making wax and honey, and the best methods for the improvement and preservation of them. . . . Translated from the French [by P. Vailant.]/London, printed for J. P. Knapton, at the Crown in Ludgate-Street; and P. Vailant, in the Strand, 1744./8 leaves, 452 p. 8 leaves (index). 12 pl./A B G H O P R S. C M W

Cole 1211, 1481.

The unsigned advertisement states 'I have taken all the materials from the memoirs of M. de Réaumur.' There is no hint in the book as to who did the translation, but it may well have been P. Vailant, who is known to have been in Paris in 1739 and may have translated from the proofs. The Vailants were a French Protestant family who settled in the Strand as foreign booksellers at the time of the Revocation of the Edict of Nantes (Plomer's *Dictionary of booksellers and printers*). There is no literal translation into English of Réaumur's *Memoires pour servir à l'histoire des insectes*, though *A short history of bees* (162), is based on them. Both books are sometimes catalogued under Réaumur.

THORLEY, Rev. John 97 1744

ΜΕΛΙΣΣΗΛΟΓΙΑ [*Melisselogia*] *or, The female monarchy.* Being an enquiry into the nature, order, and government of bees, those admirable, instructive, and useful insects. With a new, easy, and effectual method to preserve them, not only in colonies, but common hives, from that cruel death, to which their ignorant, injurious and most ingrateful owners so commonly condemn them. A secret unknown to past ages, and now published for the benefit of mankind. Written upon forty years observation and experience./London, for the author, 1744./xliii, 1 leaf, 206 p., 1 leaf, 4 pl. front./A B G H O P R S. C M W

Fig. 8. Plate 4 from the Rev. John Thorley's *Melisselogia* (1744, entry 97). It shows the author in his study; the explanation of the plates gives:

(*A*) A Swarm settling upon a Branch.

(*B*) The Bees sleeping upon a Table, and the Owner searching for the Queen. [He has stupefied the bees with the smoke from a dried fungus.]

2nd ed. *An enquiry into the nature . . .*/Front. of a new hive produced by his son, N. Thorley, and another plate of a Chinese Palace for bees./ London, Waugh, 1765./x, 158 p. 2 pl./A G R S. C M W
3rd ed./London, Johnson, 1772./Has an advertisement of Thorley's warehouse, 29 Lombard St., and of his apiary at Kingsland Green, printed over and around the front pl./x, 158 p. 2 pl./A S. M W
4th ed./1774./x, 158 p. 2 pl./Reprint of 3rd ed./B G H R. M W
Pirated eds. 1745 and 1760 (*see* 100).
Part of *Melisselogia* is reprinted in Cotton's *My bee book* (238).
Thorley's secret was the puffball narcotic, which he used when uniting colonies; its use was mentioned in 1597 by Gerard and had been long practised in the Netherlands. Of English writers, Thorley was the first to mention having found wax scales in the pockets of worker bees. The original frontispiece was a copy of Cesi's three bees—the first drawing of bees made with the aid of a microscope, 1625 (see *Bee World* 44(1): 43 (1963)). Thorley's son altered the frontispiece of the second ed. so that he could use it as a trade card, adding inter alia 'Just published the second edition of the Management of Bees by N. Thorley'. He then used the same engraving as the frontispiece of the 3rd and 4th eds. This was reprinted in 1977 by Northern Bee Books on card 30 × 21 cm.
The author is shown in his study in Fig. 8.

COLUMELLA, Lucius Junius Moderatus 98 1745
Of husbandry . . . in twelve books; translated by Michael Conrad Curtius. /London, A. Millar, 1745./xiv, 7 leaves, 600 p. 4 leaves, 39 cm./Bees pp. 385–416./A G N R S U. M W
Columella, a Spaniard, wrote about a century after Virgil. He saw service in the East and then settled on an estate near Rome. Probably a commercial beekeeper, he gave very detailed and well conceived advice on many aspects of apiary management, including requeening, uniting, and migratory beekeeping (Fraser's *Beekeeping in antiquity* (633) pp. 49–50).

ELLIS, William 99 1745
Agriculture improved, or, The practice of husbandry displayed./London, 1745./2 vol. in one./Bees Vol. 1 pp. 166–174./A O R S. W

100 1745
The female monarchy, or The natural history of bees, those admirable, instructive, and useful insects. . . . Written upon more than fifty years observation and experience; and compar'd with all the different authors who have wrote on this subject./London, printed by J. Clark, near the Temple, 1745./viii, 148 p. 2 leaves, 2 pl./O. M

2nd ed. / 1760. / a page-for-page reprint but publisher's name omitted / G. M
Chiefly a résumé of Thorley's *Melisselogia* (97). According to Walker (p. 69) these editions are probably piracies: no author is mentioned, the original matter is condensed, the religious portion omitted, and 'several conversations between Clarissa and Eugenio' added. This last was a translation from Bazin (96).

ADAMS, George (the elder) 101 1746

Micrographia illustrata, or, The knowledge of the microscope explained./ London, J. Adams, 1746. / Sting and scraper [tongue] of a bee pp. 108–110, Pl. 30. / A O

2nd ed. / 1747. / A N

4th ed. / 1747. / A O

Cole 1544.

Adams was a maker of mathematical instruments.

MAXWELL, Robert 102 1747

The practical bee-master, or, A treatise wherein the management of bees, both in common hives and in the colony way, without killing them for their honey, is, step by step, and on all probable occurrences, better and more particularly directed, than in any book hitherto published. Printed by Robert Drummond, and sold by the booksellers in Edinburgh and other towns; and at Mrs. Maxwell's shop., betwixt James's and Wardrop's Courts, in the Lawn-Market of Edinburgh, where the author may be found, or letter to him may be directed. / Edinburgh, Drummond, 1747. /viii, 138 p. / A. C M

2nd ed. / Edinburgh, J. Reid, 1750. / G O R S. M W

This ed. is also included with eds. of Reid's *The Scots gard'ner* (59).

Anr. ed. / Edinburgh, Reid, 1757. / S

Robert Maxwell of Arkland was the secretary of the Highland Society, which was dedicated to the improvement of agriculture in the Highlands. He thought the queen was a male. He described bees building comb by bringing the wax out of their mouths like swallows putting soft clay round the brim of their nest. Walker p. 47: 'An important book by a well-educated but retrograde beemaster'.

DOBBS, Arthur 103 1750

A letter . . . concerning bees and their method of gathering wax and honey./ *Philosophical Transactions of the Royal Society* 46: 536–549 (1750) / A (B) P

Dobbs wrote from Castle Dobbs in Ireland, where he had been making observations after his retirement. He comments on Réaumur and suggests that wax consists of the faeces of bees, whereas Réaumur suggested that wax was discharged from the mouth. In fact Hunter (146) was the first British author to observe the true source of wax. Dobbs did however record the correct function of the queen's spermatheca, which he had seen illustrated in Swammerdam's *Biblia naturae* (see 111). V. Grant in *Bulletin of Torrey Botancial Club* 76: 217–219 (1949) shows that, although the discovery that insect visits are necessary for the pollination of flowers has usually been attributed to J. G. Kölreuter (1761–1766), Dobbs had already made this discovery and reported it in his 1750 paper.

104 [1750]

The husbandman's jewel, directing . . . to brew ale, and beer, and make cyder, mead, mum, metheglin . . . to order bees./London, Conyers, [1750]./Of bees and how to manage them to great advantage pp. 12–14. To make mead p. 22./A

Recommended reading Worlidge 6d., Rusden 1/6d., Butler 1/– 'the only three books extant'.

MONTAGUE, Peregrine 105 [c. 1750]

Family pocket book./London, printed by Henry Coote, [c. 1750]./162 p. of which 4 are devoted to bees.

[?ARBUTHNOT, Dr. John] 106 1751

The congress of bees, or, Political remarks on the bees swarming at St. James's. From *Miscellaneous works.*/Glasgow, Carlisle, 1751./2 vol./ Bees vol. 2 pp. 130–155./A G O U

Other eds./1751, 1770, 1892.

Son of a Scottish episcopal minister who lost his preferment in 1689, Arbuthnot had to seek his fortune abroad, but in 1691 he settled in London, giving lessons in mathematics although he had a medical degree from St. Andrews. He had the good luck to be in Epsom when Prince George of Denmark was taken ill, and prescribed for him successfully. He was then appointed Physician Extraordinary to Queen Anne. Swift calls him the 'Queen's favourite physician'.

The congress of bees first appeared in the *Monthly Chronicle* on 18 July 1728. It was included in Arbuthnot's *Miscellaneous works* but is probably not by Arbuthnot; Aitken in *Life and works of John Arbuthnot* rejects it with 'tolerable confidence'.

COUNTRY GENTLEMAN 107 1753

The country gentleman's companion. / London, Author, 1753. / Of bees Vol. 1 pp. 228–234. / A N (A's copy lacks all but p. 228.)

2nd ed. / Dublin, Wilson, 1755.

Anr. ed. / London, Author, 1756. / Of bees pp. 228–234. / B

This is in fact a late issue of Markham's *A way to get wealth* (19). Vol. 1 is a reissue of *Cheape and good husbandry*, and Vol. 2 is *Country contentments*, but there is no mention of Markham or the earlier edition. The bee chapter remains unchanged.

HALE, Thomas 108 1756

A compleat body of husbandry. Containing rules for the performing in the most profitable manner, the whole business of the farmer and country gentleman . . . / London, Osborn & Shipton, 1756. / 719 p. / Of insects Book 5. Bees pp. 255–261. / A G N O R S U. W

Anr. ed. / Dublin, Wilson & Exshaw, 1757. / 4 vol. Of insects Vol. 3 pp. 142–157. / A O R

Anr. ed. compiled by John Hill, M.D., from the original papers of the late Thomas Hale, Esq. . . . and enlarged. / 1758–59. / Of bees pp. 261–72. / N R U. C M W

Hunt 556.

The work was published in 61 numbers between 28 June 1755 and 21 August 1756; see Henrey 776 and Vol. II.

Some authorities believe that Hale was an invention, and that the book was by Hill himself; see Fussell *The old English farming books*, also Hill (113).

WHITE, Rev. Stephen, M.A. 109 1756

Collateral bee-boxes, or, A new, easy, and advantageous method of managing bees: In which part of the honey is taken away, in an easy and pleasant manner, without destroying, or much disturbing the bees; early swarms, if desired, are encouraged, and late ones prevented. / London, Davis & Reymers, 1756. / 63 p. 1 pl. / A B G S. C M

2nd ed. / London, Davis & Reymers, 1759. / 1 leaf, x, 12–67 p. front. / A B G O P S U. M

3rd ed. improved. / London, Davis & Reymers, 1764. / 47 p. illus. / A G H R S. M W

Anr. ed. / London, 1766. /

White was Rector of Holton, Suffolk. Dyer in 1781 (137) and Nutt in 1832 (215) each described a similar system and claimed complete originality for it.

LISLE, Edward 110 1757
Observations in husbandry./London, J. Hughes, 1757./Bees pp. 420–421./A N S. C W
2nd ed./1757./2 vol./A G O U. W
Anr. ed./Dublin, 1757./D
Lisle (1666–1722) lived at Crux-Easten, Hants. His work was published posthumously.

SWAMMERDAM, John 111 1758
The book of nature; or, The history of insects: . . . translated from the Dutch and Latin by Thomas Flloyd. Revised and improved by notes from Réaumur and others by John Hill, M.D./London, C. G. Seyffert, 1758./xx, 4 leaves, 236 p. 153 p. lxiii, LIII pl. 6 leaves (index) 42 cm./*Treatise on the history of bees* Part I pp. 159–195./A (B) N O S. C M W
Cole 731.
Born at Amsterdam in 1637, Swammerdam studied medicine at Leyden and in 1669 published his *Biblia naturae*, a work on insects with no reference to bees. He spent 5 years studying bees, ending in 1673. He did little more work afterwards and died in 1680. A posthumous ed. published in Flemish and Latin in 1737–38 has bee matter added. The anatomical illustrations are magnificent and still considered almost faultless. Pl. XVII–XXVI depict bees. The original drawings are preserved in the University of Leyden, and the I.B.R.A. Library has photographs of them. See illustrated article by H. A. Dade, *Bee World* 53(3): 135 (1972).

BROWN, Robert 112 1759
The compleat farmer . . . containing breeding and management of horses . . . bulls . . . cows . . . sheep . . . rabbits . . . fowls . . . geese; ducks, turkies, pigeons, bees . . . making hay . . . grass, seeds, etc./London, Coote, 1759./117 p. front./Of bees pp. 85–104./A
We have seen a bookseller's catalogue giving the imprint Dublin, 1759, with the comment 'uncommon'.

HILL, John, M.D. 113 1759
The virtues of honey in preventing many of the worst disorders; to which is prefix'd an account of the origin and nature of honey, its various kinds, English and foreign; and the marks which distinguish them; also a method to obtain honey as fine in England as from any part of the world; and the best ways of taking it. With the genuine receipt for the syrup of Capillaire as made in Italy; and for the celebrated Aristaean confection.
2nd ed./London, S. Davis, 1759./54 p./A (B) S

3rd ed./London, J. Davis also M. Cooper and R. Baldwin, also J. Jackson, 1760./52 p./A M
4th ed./1761./O. M
Anr. ed./1784./.C
First edition not traced.

Hill was an apothecary and has been described as a quack doctor. He was made a Knight of the Swedish Order of Vasa for his many writings, and he wrote a prodigious number of books, including this the only book on honey before the 20th century. See *Honey: a comprehensive survey* (603). He was granted the degree of Doctor of Medicine by the University of St. Andrew's. See also Hale (108). For an account of some of Hill's other works see Henrey.

HARTE, Walter 114 1764
Essays on husbandry./London, Frederick, 1764./Bees pp. 107–111. illus./A O R S U. W
2nd ed./1770./A N R U. C
Hunt 608.

TURNER, Rev. Richard 115 1765
Plain trigonometry rendered easy and familiar calculations in arithmetick only . . . in which is also shewn a serious trigonometrical method of discovering the places where bees hive, in large woods, in order to obtain more readily the salutary produce of these little insects./London, 1765./O
Subsequent eds. to 14th./1811./A

MILLS, John, F.R.S. 116 1766
An essay on the management of bees. Wherein is shown the method of rearing those useful insects; and that the practice of saving their lives when their honey and wax are taken from them was known to the antients, and is, in itself, simple and easily executed./London, for J. Johnson and B. Davenport, 1766./xii, 159 p. illus./A G H O P R S. C M W
Anr. ed./London, 1776./S. C

This work contains nothing original, but many interesting references to English and foreign bee books, and a drawing and description of Mme. Vicat's hive. Mills was a Member of Agricultural Societies in Paris, Rouen and Berne. The frontispiece and title page are reproduced in Manley's *Honey production in the British Isles* (713), facing p. 15.

A SOCIETY OF GENTLEMEN 117 1766

The complete farmer; or, A general dictionary of husbandry in all its branches . . . together with the manner of raising bees, and of acquiring large quantities of wax and honey, without destroying these laborious insects./London, Soc. of Gentlemen, 1766./Over 200 unnumbered leaves. 35 cm. Bees occupy 14 pp./A G R S

2nd ed./1767./R U. C
3rd ed./London, 1777./G R S U
4th ed. improved & enl./London, 1793./A N S R
5th ed. rewritten and enl./1807./A N R

The account of bees is from the writings of Columella, Gedde, Warder, Thorley and Réaumur, in the form of a dictionary, with a plate showing hives.

1767

Certain ancient tracts concerning the management of landed property. See entry 1.

118 1786

COSMO, Mythogelastick, *pseud.*
(STEVENSON, John Hall)

Makarony fables with the new fable of the bees, by Mythogelastick Cosmo./London, J. Almon; also Dublin, Ewing, 1768./58 p. 27 cm./A O U

2nd ed./1768./A G U
3rd ed./1768./A
Anr. ed. as *Makarony fables* by Cosmo etc. . . . Fables for grown gentlemen./Dublin, Ewing, 1772./A

This is a political satire.

WILDMAN, Thomas 119 1768

A treatise on the management of bees; wherein is contained the natural history of those insects; with the various methods of cultivating them . . . To which are added the natural history of wasps and hornets, and the means of destroying them./London, Cadell for the author, 1768./1 leaf, xx, 170 p. 4 leaves, III pl. 27 cm./A B G N O P R S U. C M W

Anr. ed. *A treatise on the management of bees*; wherein is considered the natural history of those insects: with the manner of keeping them in hives as well as in boxes. With directions for their construction./London & Dublin, 1768./6 leaves, 78 p. index./S. M

This ed. is sometimes bound with Wildman's treatise on peach trees, with a joint title page. London, Author, also Dublin, Exshaw.

2nd ed./London, Strahan & Cadell, 1770./xvii, 1 leaf, 311 p. 4 leaves (index), appendix 16 p. III pl. 20 cm./B G S. C M

A reprint of 1st ed. plus appendix of extracts from *De re rustica*; or *The repository for select papers on agriculture, arts and manufacture.*

3rd ed./London, Strahan, 1778./xx, 318 p. 4 leaves and appendix, 16 p. III pl./A B G H P S. C W

2nd ed. reprinted Bath (Somerset), Kingsmead Reprints, 1970./B

Hunt 614.

Walker p. 78: 'Thos. and Daniel Wildman, uncle and nephew, both achieved fame by trick performances with bees which had been previously subdued by driving. Thomas, a Plymouth man, came to London in 1766, and the journals of the day report his exploits in town and country. His nephew seems to have acted independently; he gave a public exhibition at Islington (London) in 1772. Thomas was the bigger man. His treatise shows culture and an unusual knowledge of continental beekeeping, while Daniel's is decidedly inferior to it and somewhat retrograde.' Thomas Wildman's book was translated into German and Italian.

120 1768–71

Encyclopaedia Britannica. 1st ed. edited by William Smellie. Published in weekly parts to make up 3 vol. 2659 p./Edinburgh, 1768–71./A

2nd ed./1777–1783./10 vol./A. M (extract)
3rd ed./1788–1797./18 vol./A G (extract)
4th ed./1800–1810./20 vol./A. M (extract)
5th ed./1817./20 vol./A G (extract)
6th ed./1820–1823./20 vol./A
The 5th and 6th eds. were revised and corrected reprints of 4th ed.
Supplement to 4th, 5th & 6th eds./1816–1824./6 vol./A
7th ed./1830–1842./21 vol./A
8th ed./1853–1860./22 vol./A. M (extract)
9th ed./1875–1889./25 vol./A. M (extract)
10th ed./1902–1903./9th ed. + 11 supplementary vol./A
11th ed./1910–1911./29 vol./A

A facsimile reprint of the 1st ed. has been seen. An article under the title *Apis* gives an account of the 55 types of bee described by Linnaeus. Six pages deal with honeybees, with one plate and long quotations from Thomas Wildman who 'by his dexterity in the management of bees, has lately surprised the whole kingdom'. The 4th ed. has an extensive article which includes references to Maraldi, Réaumur, Debraw, Pluche, Warder, Barbut, Thorley, Swammerdam, Schirach, Wheler, White, and 'a true account' of the origin of honeydew by the Abbé Boissier de Sauvages (23 p. and 2 pl.). Walker p. 26 states that the supplement to 4th, 5th and

6th eds. contains 22 pages on bees. The 7th and 8th eds. have major articles by James Wilson, and the 9th ed. by John Hunter, who was at one time Secretary of British Bee-Keepers' Association. Wilson's article on Entomology was later reprinted under the title *A treatise on insects.*
The 11th ed. had articles *Bee* by G. H. Carpenter and *Bee-keeping* by William Broughton Carr.
Later American editions of the *Encyclopaedia Britannica* are outside the scope of this *Bibliography.*

PETRIE, James 121 1769
The Scots' apiary, or *The compleat bee-master.* Unfolding the whole art and mystery of managing bees: wherein is shewn, the invaluable secret of obtaining their labours and preserving their lives. To which is added, the cultivation of hopps and lyquorice root, both of which the author brought to great perfection in Galloway . . . The like never before published. With an appendix containing some remarkable occurrences during the author's life./Printed for, and to be sold by the Author (in the eighty-third year of his life), Glasgow, 1769./44 p./S

List of subscribers, accounting for 164 copies.

[WESTON, Richard] 122 1769
Tracts on practical agriculture and gardening . . . particularly addressed to the gentlemen farmers in Great Britain . . . to which is added a complete chronological catalogue of English authors on agriculture, gardening, &c. By a country gentleman./London, Hooper, 1769./xxiii, 277 p. 70 p. 10 leaves./A O R. C W

2nd ed./London, Hooper, 1773./A N O R
3rd ed./Leicester, Pares, [n.d.]

Henrey 1435, 1490–91./Hunt 605.

The tracts contain nothing on bees, but the 70-page catalogue, with its title and author index, contains many references to bee books. The title page of the 2nd ed. is reproduced in Fussell.

WHITE, William 123 [1771]
A complete guide to the mystery and management of bees, containing instructions how to manage them with respect to their breeding, gathering, swarming, hiving, feeding etc. to considerable advantage: also directions whereby the governor or queen-bee may be distinctly known; together with several curious matters concerning them, deduced chiefly from experience. Being the most valuable discovery hitherto found out, by W. White Sen. of Shutford, near Banbury, Oxon. who, in the year 1766, was honoured with a premium, by the Society of Arts and Sciences, for

his singular abilities therein./London, for the Author, [1771]./xvi, 94 p. 2 leaves, engraved front./A B G O R S. C M W

Anr. ed. To which is added, *A practical monthly bee calendar*, by the late James Beesley, of Banbury./London, Simpkin, Marshall, also Hamilton, Adams, also Oxford, H. Slatter, also Reading (Berks.), Rusher & Johnson, also Banbury (Oxon.), J. G. Rusher./1852./A N O S. M

The peculiar type used in this 1852 ed. had been patented by Rusher, the printer, the idea being that more matter could be got into a page by eliminating the tails (descenders) of the letters.

Facsimile reprint of 1852 ed./Mytholmroyd (Yorks.), Northern Bee Books, [1977]./viii, 2 leaves, 14–75 p./B

COOKE, George 124 1772

The compleat English farmer, or, Husbandry made perfectly easy in all its branches. Containing what every farmer ought to know and practice. Among the various Articles treated of in this work, are the following, viz . . . The Management of Bees both for profit and pleasure./Boston (Lincs.), D. Kneeland for Henry Knox, 1772./Bees pp. 86–97./A G O R

Anr. ed. 1775./A

There has been some uncertainty about the date, but the Bodleian Library informed us: 'Our copy can be dated 1772 by a "This day is published" advertisement of *Companion for the fireside*, which bears that date on its title.' The title page is reproduced in Fussell's *More old English farming books 1731–1793*.

WILDMAN, Daniel 125 1773

A complete guide for the management of bees throughout the year . . . containing 1. A description of the queen bee. 2. The generation of bees. 3. Of the drones. 4. Of the proper situation for a bee-house. 5. The proper method of swarming and hiving. 6. Of separating the honey and wax. 7. Of the enemies and sickness incident to bees. 8. Of feeding them in the winter season. 9. Explanations of the newly-invented hives, with proper directions in what manner they are to be made use of./London, printed for the Author and sold by him at his house Gray's-Inn Gate, Holborn, 1773./39 p. 1 pl.

Anr. ed./London, Author, 1775/A C

Anr. ed./London, Author, 1778./39 p./M

2nd ed./London, Author, 1779./48 p. 2 pl./A O P. M

3rd ed. with additions/London, printed for the Author, and sold by him at his bee and honey warehouse, No. 326, Holborn; where any quantity of his new-invented hives may be had on the shortest notice. 1780./xiii, 9–48 p. 2 pl./B G R S. C M W

Subsequent editions followed quickly: 4th and 5th 1785, 6th 1788, 7th 1789, 11th, 12th and 13th 1792, 14th 1799, 15th 1801, 16th 1802, 17th 1808, 18th 1809, 19th 1812, 20th 1819./A B (14th ed.) G H P R S. C M W Not all these editions have been seen, but they are believed to be the same, with 48 p. and 2 pl. Dates of 8th, 9th and 10th eds. not known. Abstracted and reprinted 1844 and 1851, see Munn's *A description of a bar and frame hive* (250).

For a note on Daniel Wildman see under Thomas Wildman (119) and Neighbour (307). Daniel's book was clearly a success, as the number of editions shows. We have not seen 1st ed. In 1774 one French translation appeared in Paris, and another in Amsterdam. The book was also translated into German and Italian, Johansson's entry 562 says the text of the main part of *A complete guide . . .* by a farmer of Massachusetts, published by L. Thomas in 1792, was taken from Daniel Wildman's book.

BRYANT, Jacob 126 1774

A new system, or, An analysis of ancient mythology wherein an attempt is made to divest tradition of fable; and to reduce the truth to its original purity; . . ./London, T. Payne, 1774–76./3 vol./A O U

2nd ed./1775–76./3 vol./D

3rd ed./London, J. Walker, 1807./6 vol./A O U

GOLDSMITH, Oliver 127 1774

A history of the earth and animated nature./London, J. Nourse, 1774./ Bees Vol. 8 pp. 65–95./A

Subsequent eds. 1779, 1805, 1808, 1810, 1824, 1837 etc.

Cole 1920.

Bee matter taken from Réaumur. Goldsmith, poet and playwright, was the author of many works including a periodical *The bee: being essays on the most interesting subjects;* this was, however, not about bees. Only 8 issues appeared, which were published together in book form in December 1759.

RINGSTED, Josiah 128 [c. 1775]

The farmer: comprehending . . . the ordering of bees and their labours./ London, Dixwell, [c. 1775]./172 p./Bees pp. 126–134./R U. M

Subsequent eds. 1796, 1800./G O R. W

See *The farmer's wife* (134).

HARRIS, Moses 129 1776

An exposition of English insects with curious observations and remarks, wherein each insect is particularly described; its parts and properties considered; the different sexes distinguished, and the natural history faithfully related./London, Author, 1776./Bees pp. 130–138. illus./A D U. W

Anr. ed./1781./A

Anr. ed./London, 1782./illus./A O S. C W

Lisney 239–243.

DEBRAW, J. 130 1777

Discoveries on the sex of bees . . . Philosophical Transactions of the Royal Society 67: 15–32 (1777). Communicated to the Society by Rev. Nevil Maskelyne, B.D., F.R.S., and Astronomer Royal, and read 21 November 1776./A (B) P

Debraw asserted that the drones fertilized the eggs in the hive after they had been laid, rather in the manner of fishes; he also said he had observed drones smaller than the usual drones. He confirmed Schirach's discovery that bees can raise queens from the eggs of workers, saying that he had started his experiments nearly two years before Schirach's publication. The paper attracted great attention. The following year Nathaniel Polhill, M.P. for Southwark, published a letter to John Belchier, F.R.S., confirming Debraw's results and saying that he was intending to experiment with raising queens 'with so little danger from the stings of the bees, as may recommend it to little farmers and cottagers in the country'. *Phil. Trans.* 68: 107–110 (1778) read 22 January 1778./A P

Debraw's paper was also reprinted as an appendix to Vol. 2 of *Dissertations relative to the natural history of animals and vegetables*, translated from the Italian of Spallanzani./London, J. Murray, 1784./pp. 357–372./A

EDITORS OF FARMER'S MAGAZINE 131 1779

General dictionary of husbandry, planting and gardening . . . selected from the best authorities./Bath (Somerset), printed by R. Cruttwell and sold by W. Goldsmith . . . London, 1779./In 2 vol./A

The 2 vol. contain over 400 unnumbered leaves. The article *Bee* occupies 6 pp.

Henrey 742.

Compiled in the manner of a true dictionary. Henrey states that some authorities ascribe the work to Cruttwell, but that she can find no supporting evidence. She quotes advertisements of 1786 and 1797 describing the work as by 'C. Williamson and others'.

THOMPSON, William 132 [1779]

The new gardener's calendar, or, Every man a complete gardener . . . likewise a new essay on the mystery and management of bees, in which the best instructions are given with regard to their breeding, gathering, swarming, hiving, etc./London, Hogg, [1779]./Useful instructions for the management of bees pp. 107–115./A O. W

Henrey 1428.

Thompson was a professional gardener.

COOKE, Samuel 133 [c. 1780]

The complete English gardener: or, Gardening made perfectly easy . . . to which is added *The complete bee-master*, or, Best method for managing bees, both for profit and pleasure./London, Cooke, [c. 1780]./Bees pp. 97–118./A R S. W

2nd ed./London, Cooke, [n.d.]./157 p./Bees pp. 86–106./A O S. M

Henrey 575–577.

There is some uncertainty about the date of this book. But Henrey gives c. 1780, also Walker, and Fraser's *History* p. 58 says 'Each of the two copies I have seen contained the signature of the first owner and the date 1780.'

Cooke was gardener at Overton in Wiltshire. The book contains a short account of skep beekeeping, and recommends killing the bees rather than driving them to take the honey.

134 [1780]

The farmer's wife, or, Complete country housewife . . . containing . . . the method of making cyder, perry, mead, . . . ample directions respecting the management of bees, with an account of the use of honey./London, A. Hogg, [1780]./Bees pp. 96–108. Mead p. 50./A G N

Fraser's *History* p. 55 says the wording of the 12 pages on bees is identical with that in *The Farmer* by Ringsted (128).

KEYS, John 135 [1780]

The practical bee-master in which will be shewn how to manage bees either in straw hives or in boxes, without destroying them, and with more ease, safety and profit, than by any method hitherto made public . . . interspersed with occasional strictures on Mr. Thomas Wildman's *Treatise on bees* . . ./London, printed for the Author, and sold by him at his house in Cheshunt-Street, Hertfordshire; J. Johnson, No. 72, St. Paul's Church-Yard, nr. Vallance, Cheapside; and by principal booksellers in town and country, [1780]./2 leaves, iii–x, 390 p. 1 leaf, 1 pl. (pull-out)/A B G H N O P R S U

Later copies have an additional leaf with further errata and a postscript.

BARBUT, James 136 1781

The genera insectorum of Linnaeus exemplified by the various specimens of English insects drawn from nature (In English and French)./London, Sewell, 1781./Genus VIII *Apes*—Bees pp. 268–275, Pl. 15./A

Lisney 289./Cole 1963.

'The office of the males or drones is to render the queen pregnant.'

[DYER, W.] 137 1781

The apiary laid open, or, An advantageous treatise on the construction and use of retrocoupling bee boxes (a new invention). Also plain instruction for the management of bees in general . . . their preservation from cold, heat, enemies, and hunger . . . the making a swarm by art, never treated of before . . . Deduced from 30 years' experience by W.D., of Shipton-Oliffe, Glos. / Tewkesbury (Glos.), for Author, 1781. / 191 p. / M

Walker p. 19: 'The only other copy I have seen of this very scarce book is in the library of the Lond[on] Entomological Soc[iety]. On pp. 7 and 157, the author refers to 'figures on the frontispiece'. As, however, this is wanting in both copies . . . it is probable that none was published.' Similar collateral boxes had been recommended previously by S. White (109) and were later advocated by Nutt (215) and J. Jones (246).

CRÈVECOEUR, Hector St. John de 137a 1782

Letters from an American farmer./London, Davies & Davis, 1782./318 p./A

Letter II: 'My bees, above any other tenants of my farm attract my attention and respect.' Crèvecoeur tells how he shot a king bird and took 171 bees from his craw. He laid them in the sun and 54 of them revived. If a swarm flies into the woods it does not matter, because in the autumn Crèvecoeur finds the nests by tracking marked bees.

Various eds. Published also in Dublin 1782 and Belfast 1783.
A French version published in Paris 1784.

Selection, with wood engraving of bee hunting, entitled *The divided loyalist.*/London, Folio Society, 1978./A B

BROMWICH, Bryan J'Anson 138 1783

The experienced beekeeper: containing an essay on the management of bees; wherein is shewn, from long practice, the most easy and profitable method of treating those useful insects . . . particularly interesting to the keeper of bees, and useful to every family. Together with an improved method of making mead, and other wines, with honey./London, Dilly, 1783./xiv, 66 p. 1 pl./A G P S. C W

2nd ed./London, Dilly, 1783./xiv, 66 p. 1 pl./A B G H O S. M

Walker p. 10: 'Except for the added words, '2nd edn.', the 2nd precisely resembles [the 1st].'

Fraser's *History* p. 60 states: 'He was the first to discover the possibility of keeping more than one queen in a hive: three straw hives blew over in a high wind, and fell in such a position that the bees and their queens joined forces in one hive. As a result of this discovery, Bromwich experimented with collateral boxes . . .'

ADAMS, George (the younger) 139 1787

Essays on the microscope./London, R. Hindmarsh, for the Author, 1787./Bees pp. 257–265./A D G O U

2nd ed. With considerable additions and improvements by Frederick Kanmacher . . ./London, for the editor and for W. & S. Jones, 1798./ 27 cm. *Of the apis or bee* pp. 279–285./A B N O R. C

Cole 2019–2020.

Adams was a mathematical instrument maker. He knew of Schirach's discovery that bees could raise a queen from a worker egg.

HOY, Richard 140 1788

Proper directions how to manage bees in Hoy's octagonal box bee-hives. London, issued by R. Hoy from his honey-warehouse at 175 Piccadilly, 1788./4 p. illus./B

Milton's *Practical bee-keeper* (248), 1843 edition p. 145, lists Hoy as a writer on bees against the date 1782, but the only known copy of this broadsheet is dated 1788. The engraving is reproduced in *Honey: a comprehensive survey* (603).

MARSHALL, William 141 1788

The rural economy of Yorkshire comprising the management of landed estates and the present practice of husbandry in the agricultural districts of that county./London, Cadell, 1788./Bees Vol. 2. pp. 238–244./A N O R. W

2nd ed./London, Cadell, 1796./Bees Vol. 2. pp. 238–243./D O R. W

BONNER, James 142 1789

The bee-master's companion and assistant wherein is set forth the properest methods of managing those insects, so as they may turn out to the best advantage . . . also how to make all your hives equal in bees, so as never to have any weak hive; with an account of the power the working bees are invested with of raising any egg in the hive to be a queen when the community stands in need of one./Berwick, J. Taylor for the Author, 1789./xii, 225 p./A G H S. M W

WHITE, Rev. Gilbert 143 1789
The natural history of Selbourne, with observations on various parts of nature, and the naturalist's calendar./London, D. White, 1789./Letter XXVII, 12 December 1775, is about an idiot boy who ate bees, pp. 200–201./A
Many editions.
Hunt 707.

SMELLIE, William 144 1790
The philosophy of natural history./Edinburgh, Bell & Bradfute, 1790–1799./Vol. 1 Bees pp. 336–349 etc. Vol. 2. Language of bees p. 433./A N O
Subsequent eds. 1808. 1829./. C W
Cole 2048.

HERON, Robert 145 1792
Elegant extracts of natural history: collected and revised from a great variety of the most elegant and authentic writers in this science./Edinburgh and London, Balfour, Robinson & Dilly, 1792./Bees pp. 166–213./A M

HUNTER, John, F.R.S. 146 1792
Observations on bees./Philosophical Transactions of the Royal Society 82: 128–195 (1792) read 23 February 1792./A (B) P. M
This is an important and much quoted paper based on Hunter's own observations. Hunter observed that wax was secreted between the scales of the body. In fact this had been reported by Hornbostel in 1744, and Thorley (97) had found wax scales on the belly of a bee 'within the Plaits' but seems to have thought that wax was collected by the bees. Fraser's *History* p. 64 says that Hunter rediscovered the dancing of the bees, which as far as he knew had not been recorded since the time of Aristotle. In fact Evelyn (34) had observed bees dancing on the surface of a swarm and thought it was a method of communicating a favourable site for the new colony—'the out guards that by a kind of shivering motion communicate it to the whole swarm'. Hunter thought the bees were shaking out the cells of wax from their bodies, 'but I could not absolutely determine what they did'.
Reprinted in 1806 (*see* 172).

147 1792

The natural history of insects. Compiled from Swammerdam, Brookes, Goldsmith, etc./Perth, R. Morison and others, 1792./2 leaves, 210 p. 20 pl./Bees pp. 89–119, Pl. 14/A O S

Anr. ed./London, Law and others, 1792./
Anr. ed./1800.
Anr. ed./[1814].

Lisney 346–350./Cole 2061. *The National Union Catalog* lists 8 copies in U.S.A.

SYDSERFF, Robert 148 1792

Sydserff's treatise on bees; being the result of upwards of thirty years experience. Executed on a more general, extensive, familiar, and useful plan, than any other work of the kind yet published./Salisbury (Wilts.), Collins, 1792./xi, 99 p. illus./A G H P S. C M

Sydserff's book is practical and amusing. Walker p. 67: '[he was] b[orn] at Stalbridge in Dorsetshire [and] d[ied] at Mells [*sic*], Somerset. [He was] by trade a tailor, [and] by calling a Methodist preacher.' Part of his book is reprinted in Cotton's *My bee book* (238).

ROBERTSON, George 149 1793

General view of the agriculture of Midlothian./Edinburgh, Nicol, 1793./Bees p. 163 and Appendix No. 10 pp. 85–96./A O
Anr. ed./1795./A S

MONK, John 150 1794

An agricultural dictionary consisting of extracts from the most celebrated authors and papers./London, G. Woodfall, 1794./Bees pp. 58–59./A

Section by John Keys also published in *Letters and papers on agriculture, &c., addressed to the Bath Agricultural Society* 5: 320 (1790), in which Keys urges the Society to investigate Schirach's discovery that a queen could be raised from a worker larva, because he had tried to achieve this and failed.

BONNER, James 151 1795

A new plan for speedily increasing the number of bee-hives in Scotland; and which may be extended with equal success, to England, Ireland, America, or to any other part of the world capable of producing flowers./Edinburgh, J. Moir and others, 1795./xx, 258 p./A B G H O P R S. C M W

B's copy has 'Printed and sold by the author and by Brash & Reid, Glasgow', and 'world' is printed 'word'.

Anr. ed. *A treatise on the new natural history and management of bees*, with a new plan founded on practice./Edinburgh, J. Moir, 1796./xx, 258 p./G. M

Walker p. 8: '[Printed] from [the] same type as [the] 1795 edn.; the title alone being altered.'—the name being spelt Bonar.

Bonner who was Bee-Master at Auchencrow, near Berwick-on-Tweed, was the most able and the best known of Scottish bee-men.

BONNER, James 152 1795

On culture of bees: a letter addressed to Sir John Sinclair. In *General view of the agriculture of the northern counties and islands of Scotland.*/London, Mackie, 1795./Appendix 2, pp. 13–15./(S)

VIRGIL (Publius Vergilius Maro) 153 1795

The episode of Aristaeus; translated from the fourth book of the Georgics of Virgil by Percy Clinton Smythe./Dublin, Jackson, 1795./20 p./A

KEYS, John 154 1796

The antient bee-master's farewell; or, Full and plain directions for the management of bees to the greatest advantage; disclosing further improvements of the hives, boxes, and other instruments, to facilitate the operations; especially that of separating double and treble hives or boxes, with certainty and safety, without injuring the bees. . . . interspersed with new but important observations. Also brief remarks on Schirach and other distinguished Apiators on the Continent. Deduced from a series of experiments during thirty years./London, Robinson, 1796./xvi, 273 p, II pl./A B G H P R S. C M W

Anr. ed. Title as above but no author named./Dublin, Byrne, 1796./ 252 p. II pl./.M

Anr. ed./*A treatise on the breeding and management of bees*, to the greatest advantage. Interspersed with important observations, adapted to general use. Deduced from a series of experiments during thirty years. A new edition./London, Lackington (and others), 1814./2 leaves, xvi, 272 p. 2 leaves, II pl./A B G R S. M

Although described as a new edition, the text (including the Preface) and the plates are the same as in the 1796 ed. Keys would not accept Schirach's discovery that a queen could be raised from a worker larva; see Monk (150).

[LETTSOM, John Coakley] 155 1796?

Hints for promoting a bee society./London, Darton & Harvey, 1796?./ 8 p. 1 pl./

2nd ed./London, Darton & Harvey, 1796./22 p. 1 pl. 1 leaf./A G. M

Southampton University Library holds a copy of the first edition which lacks the drawing of a bee hive mentioned in the text, and has no title page. The 2nd ed. is dated 1796.

Lettsom joined the Western Apiarian Society in 1799 and frequently wrote for the *Gentleman's Magazine*, which published a review of his pamphlet, 66–1033. It is also mentioned in the *Transactions of the Western Apiarian Society* (158). Printed *Rules of the Ulverston Apiarian Society* (4 p. undated) were found inserted in G's copy of Warder's *True Amazons* (74), 8th ed.

ISAAC, Jacob 156 1799

The general apiarian wherein a simple, humane and advantageous method of obtaining the produce of bees without destroying them, is pointed out in a series of letters to a friend./Exeter (Devon), Trewman, 1799./108 p. 1 pl./A G H S. C M W

Anr. ed./Exeter (Devon), Trewman, 1803./128 p. illus. 1 pl./A B G P R S. M W

Isaac was Secretary to the Western Apiarian Society, founded in 1799, to whom this book is inscribed. Both eds. show a skep on the title page, and have a plate showing a bee house. The 1803 ed. has a description of the hive invented by Mr. James Roberts of Crediton, Devon, and an illustration of it on p. 113.

VANIÈRE, Jacques 157 1799

The bees, a poem; translated by A. Murphy from the 14th book of *Praedium rusticum*. (Paris, 1696)./London, Rivington, 1799./70 p./A G O S. M

Walker p. 18: 'Murphy [was] author, actor, and barrister . . . a close friend of Dr. Sam[uel] Johnson and intimate with his . . . patron Thrale, to one of whose daughters . . . the poem is dedicated.'

WESTERN APIARIAN SOCIETY 158 1799–1809

Rules and Transactions./Exeter (Devon), Trewman./A S. M.

Rules first printed 1799./300 copies of revised rules printed 1807./18 p. *Transactions* published in 7 parts, more or less annually, 1800–1809./ Altogether 212 p.

Walker p. 75: 'The only other complete copy known to me is in [the] B[ritish] M[useum].' In fact the Dr. Williams Library, Gordon Square, London, has a copy presented by Jacob Isaac, whose book *The general apiarian* (156) is bound with the 1807 *Rules* and the 7 parts of the *Transactions*. A manuscript note by Isaac states that 'the Society has been dissolved.'

1800 to 1849

Preserving the bees and improving the lot of the beekeeper

159 1800

Comfort to Aristeus, or, A few useful hints in the management of bees, so as to render honey and wax a cheap and plentiful commodity, preserve the lives of these useful animals, and open a new source of industry and wealth to the gentlemen and farmers, new and profitable. / London, printed for the Author and sold at No. 90 Edgware Road, Paddington, 1800. / 30 p.

The only copies known are in the libraries of the Royal Agricultural Society and the Royal Entomological Society.

160 [c. 1800]

Encyclopaedia Perthensis. / [c. 1800]. / Bee Vol. 3 pp. 461–490. illus. / .M W

2nd ed. / Edinburgh, J. Brown, 1816. / Bee Vol. 3 pp. 461–490. illus. / A O . M

Walker p. 21: 'A far ranging article, with good quotations.'

P., L. 161 [1800]

The life of a bee, related by herself. / London, Marshall, [1800]. / xi, 71 p. front. 15 cm / A

For children. / Hodgson 434.

SCHULZE, Christian 161a 1800

A memoir on the great advantage of raising silkworms and of cultivating bees in Ireland in a letter to the right hon. & honrable [*sic*] The Dublin Society. / *Transactions of the Dublin Society* 1: 75–96 (1800). A

162 1800

A short history of bees. In two parts. I. The natural history of bees, with directions for the management of them, an account of their enemies, &c. from Réaumur, &c. II. An ænigmatical account of a neighbouring nation —their queen, her palaces, attendants, &c. / London, Vernor & Hood, 1800. / xxiii, 133 p. 3 leaves, front. / A B G R S . C M

Pub. in U.S.A. 1803. / Johansson 4.

Page 88: 'The mattras on which the royal infant is placed, is of a thicker substance than what is used for common babes. This mattras is at once a bed and food.'

VARRO, M. Terentius 163 1800

The three books of M. Terentius Varro concerning agriculture. Translated by the Rev. T. Owen, M.A./Oxford, at the University Press for the Author, 1800./xi, 257 p. 1 leaf./Bees pp. 239–253./A B

Owen was Rector of Upton Scudamore (Wilts.). He also translated Cato's *Agriculture* (1803), and B's copy contains the two works bound together. *See also* 169.

164 1802

The domestic encyclopaedia Vol. 1./London, 1802./Bee pp. 214–232./. M

Walker p. 22: 'Contains transl[ation] of an account in 'Atti della Soc. Patriot. di Milano,' 1789, of Prof. Gaetano-Harasti's bee-hive, which attracted much notice in Italy, and was described in 'Univ. Mag.', Lond[on], Dec., 1799.'

KIRBY, William 165 1802

Monographia apum angliae, or, An attempt to divide into their natural genera and families, such species of the Linnean genus Apis, as have been discovered in England . . ./Ipswich (Suffolk), printed for the Author by J. Raw, and sold by J. White, Fleet-Street, London, 1802./Vol. I 1 leaf, xxii, 258 p. 14 pl. Vol. II 388 p. 18 pl. (some col.)/A B G O R S . C M W

A SOCIETY OF PRACTICAL GARDENERS 166 1802

Rural recreations, or, The gardener's instructor. With a treatise on the management of bees./London, Vernor & Hood, 1802./Bees pp. 220–232, 278–288, 324–336, 370–379. illus./A P R S . M

PLAYFAIR, James 167 1804

On the care and knowledge of bees. Unpublished manuscript, 1804./290 p. /S

See article by J. N. Tennent, 'On the care and knowledge of bees'. *Scottish Beekeeper* 27 (12):236–237 (1951).

In an article on the manuscript in *Scottish Beekeeper* 27(12):236–237 (1951), Dr. J. N. Tennent suggests that Playfair's rejection of Huber's statements on the fecundation of the queen by the drone may have led to his difficulty in finding a publisher. Playfair describes and illustrates an almost cubical wooden hive, the sides being 12 inches at the top and 11½ inches at the bottom. It had eight frames, but he did not fully recognize the importance of the bee space.

BINGLEY, Rev. William 168 1804

Animal biography, or, Authentic anecdotes of the lives, manners, and economy of the animal creation according to the system of Linnaeus.

2nd ed. with considerable additions, in 3 vol./London, Phillips, 1804./ The bee tribe Vol. 3 pp. 375–402./A

Subsequent eds./1805, 1813, 1820./AS

169 1805

Geoponica—agricultural pursuits; translated from the Greek by the Rev. T. Owen, M.A./London, W. Spilsbury for the Author, 1805./ Preparation of hydromel Book 8 pp. 257–258./A

'Others indeed mix snow only with honey and having well wrought it, they lay it by, and it is a medecine used in raging fevers.'
See also 163.

REES, Abraham 169a [1805]

The cyclopaedia; or, Universal dictionary of arts, sciences and literature./ London, Longman & others, 1805./In 39 vols., articles arranged alphabetically but not paged. 27 cm./A B (bee article). M (bee article)

Contains a long article entitled *Bee*, also one entitled *Hive*. The former is in Vol. IV pub. in 1805, although the volume is usually found with a title page dated 1819 which was issued when the work was complete.

EVANS, John 170 1806

The bees: a poem in four books. With notes, moral, political and philosophical./Shrewsbury (Salop.), J. & W. Eddowes./Book 1 1806. xi, 79 p. Book 2 1808. 95 p. Book 3 1813. 123 p. Book 4 was never published. 27 cm./A B (1, 2 only) G S . C M

Walker p. 23: '[Evans (1755–1846)] was senior physician to the Salop Infirmary, Shrewsbury; he founded the public library and wrote many papers on medical and botanical subjects. 'The Bees' is by far the most ambitious and learned poem on the subject in the English language. The first two Books were reprinted in *B.B.J.*, 1876–77.' The poem was quoted by Bevan in 1827 (201), but its chief enthusiast was Watts, who in 1880 edited the 7th ed. of Taylor (227).

HUBER, Francis 171 1806

New observations on the natural history of bees . . . translated from the original [by Sir J. G. Dalzell]./Edinburgh, Anderson, 1806./viii, 300 p. 2 pl. in one./A G S. C M W

2nd ed./Edinburgh, Anderson, 1808./xxv, 314 p. 2 pl. in one./B G O P R S. C M W

3rd ed./Edinburgh, Tait, 1821./xv, 440 p. 1 leaf, V pl./A B D G H N O S. C M W

New ed./*Observations on the natural history of bees.*/Cupar (Fife), Tullis, 1840./xxiv, 352 p. V pl./S. M

New ed./*Huber's natural history of the honey bee.*/London, Tegg, 1841./ xxiv, 352 p. V pl. B G O S. M W/Some copies have both the 1840 and 1841 title pages.

Cole (French ed. 1792) 2065.

François Huber's *Nouvelles observations sur les abeilles* was first published in Geneva in 1792. A further edition was published in Paris in 1814 in two volumes with additional plates. This edition incorporated a large amount of additional material, much of which had appeared in the meantime in the *Bibliothèque Britannique*, and was edited in part by Huber's son Pierre. The 1821 and 1840 English translations are somewhat abbreviated, and the first full English translation was that published in U.S.A. in 1926, by C. P. Dadant (Johansson 238).

SHAW, Geo. 172 1806

General zoology./London, Kearsley, 1806./Bee Vol. 6 pp. 289–349. illus./A O . C M

Includes a reprint of the important paper by J. Hunter on the secretion of wax; *see* 146.

DICKSON, R. W. 173 1807

Practical agriculture, or, A complete system of modern husbandry . . . with the management of live stock./London, Phillips, 1807./Bee stands Vol. 1 p. 68./Bees Vol. 2 pp. 1217–1221./A

Quotes Keys, Bonner, and Marshall.

PARKINSON, Richard 174 1810

Treatise on the breeding and management of live stock, comprising cattle, sheep, . . . bees./London, Cadell & Davies, 1810./Bees Vol. 2 pp. 389–408./A N O R . C

Includes construction of a bee house 13 ft long, 4 ft high and 14 inches wide. 'Hives are made of straw. To preserve them, it has been my custom to cover the top with an earthen milk vessel, turned upside down . . .'

SANDHAM, Miss Elizabeth 175 1812

The perambulations of a bee and a butterfly./London, Tabard, 1812./A

Subsequent eds./1816, 1819./132 p./A G S . M

A moral tale for children./Hodgson 426.

CUTWODE, T. 1815
Caltha poetarum. (*see* 15)

HUISH, Robert 176 1815
A treatise on the nature, economy, and practical management of bees; in which the various systems of the British and foreign apiarians are examined, and the most improved methods laid down for effectually preserving the lives of the bees. Containing also an accurate description of the hives invented by Lombard, Ducouedic, Huber, Vicat, L'abbé della Rocca, and other foreign apiarians; and of the newly invented hive for depriving the bees of their honey, with safety and expedition. Forming the most complete guide to the study and management of those valuable insects./London, Baldwin, Cradock & Joy, 1815./xxiii, 414 p. 6 pl./ A B G H N O R S . C M W

Dedicated to Sir Joseph Banks, G.C.B.

2nd ed. with additions./London, Baldwin, Cradock & Joy, 1817./xxxvii 1 leaf, 396 p. 6 pl./A B G H P S . C M W

Dedicated to Sir Thomas Clarges, Bart. and George Isaac Call, Esq. of Taplow Hill, near Maidenhead.

New ed. enl. *Bees: their natural history and general management*; comprising a full and experimental examination of the various systems of native and foreign apiarians./London, Sherwood, Gilbert & Piper, 1842./458 p. illus./E G

New ed. greatly enl./London, Bohn, 1844./xxvii, 9–458 p. illus. front. (portrait of author)./B G S . M W

Huish (1777–1850) attacked some of 'the absurdities of M. Huber', and the New ed. contains an analytical exposition of the errors of the theory of Huber. Huish wrote many other works, including accounts of trials, and romances entitled *The brothers; or The castle of Niolo* and *The nun of Gnadenzell*, and also moral tales in *Our grandmama's clock* and *Our grandpapa's chest*. He describes himself as 'Author of *The Peruvians*, a poem &c. &c.'

KIRBY, William & SPENCE, William 177 1815
An introduction to entomology: or elements of the natural history of insects./London, Longman Hurst and others, 1815–1826./IV Vol., XXX pl. (5 col.)/A G S

The work was completed in 1826 by which time the first two volumes had both reached a 4th ed. Most sets are therefore of mixed editions, and sometimes the plates are bound as a 5th vol./5th ed. 4 vol. 1828./6th ed.

Vol. I and II 1843./7th ed. Vol. I and II in one vol. no pl. 1856./Later eds. 1858, 1859, 1865, 1870, 1885./A B (7th & 8th eds.) G K S

Published in U.S.A. 1846./Johansson 274.

There were 750 copies of the 1st edition of the 1st volume in 1815. It was immediately successful, and later editions were produced in 1816 and 1817 when the 2nd volume appeared. There were further editions of the 2nd volume in 1818 and 1822.
Bees are mentioned in all 4 volumes but appear mainly in vol. II pp. 116–217, Letters XIX and XX, containing an account taken principally from Réaumur and Huber but with numerous references to other authors.
The book is written in a series of letters, the better to allow latitude for amusing digressions, thereby making it more palatable for the reader.

FRANKLY, Mr. 178 1816

The history of Frugal, the wild bee./London, Whittingham & Arliss, 1816. /107 p. col. front. 14 cm./A B O S . M

A quaint and strictly moral children's book./Hodgson 421.

TAYLOR, Joseph 179 1817

Anecdotes of remarkable insects; selected from natural history, and interspersed with poetry./London, Baldwin, 1817./3 leaves, 224 p./Bees pp. 61–96./A

For children./Hodgson 419.

BRITISH APIARIAN SOCIETY 180 1819

Rules of the British Apiarian Society: established for the promotion of the culture of the bee amongst the cottagers./Maidenhead (Berks.), G. W. Wilton, 1819./16 p./A.M

Walker's copy was (p. 27) typed from the original in the Drory Library, Berlin. The Society was founded by Huish, who was for some years its secretary. He said in his *Treatise* (*see* 176): 'The establishment of an apiarian society was a few years ago suggested by myself . . . The particulars of the plan will be transmitted to any gentleman who may be willing to second the intentions of the author in the formation of a society.'

181 1819

Dialogues on entomology in which the forms and habits of insects are familiarly explained, illustrated with twenty-five engravings./London, R. Hunter, 1819./408 p. Bees pp. 252–267. Wild bees pp. 268–281./A

Written in the form of dialogues between Lucy and Mother.

ESPINASSE, Isaac 182 1819

Management of bees./*Transactions of the Royal Society of Arts* 36: 29–52 (1819)./illus./B

The author was voted the Silver Ceres Medal of the Society for this communication, headed Chancery Lane, Feb. 7, 1818.

HUISH, Robert 183 [1819]

Instructions for using the Huish hive, from which the combs are extracted without killing the bees./London, Wetton & Jarvis for British Apiarian Society, [1819]./26 p./A S. M

2nd ed./1822./M

Copies of *The cottager's manual* (184) dated 1821 describe the 2nd ed. of the *Instructions* as 'just published'.

Huish weighed 100 bees in summer and winter.

HUISH, Robert 184 1820

The cottager's manual for the management of his bees, for every month in the year; both on the suffocating and depriving system./London, Wetton & Jarvis, 1820./1 leaf, xviii, 84 p./A G S . M

2nd ed. . . . with the necessary instructions for the purchase of hives, and the general management of the apiary with considerable additions./London, Wetton & Jarvis, [1821]./xvi, 17–99 p./A B G . M W
The date is on the outside cover but not on the title page.
Anr. ed. enl./Maidenhead (Berks.) & London, Wetton & Jarvis, 1832./104 p./A G S . M W

185 1820

Family receipt book, or universal repository of useful knowledge and experience in all the various branches of domestic economy . . ./London, for Editors, 1820./Hydromel, mead, metheglin pp. 339–348./A

BUCKE, Charles 186 1821

Beauties, harmonies and sublimities of nature with occasional remarks on the laws, customs, manners and opinions of various nations./London, G. & W. B. Whittaker, 1821./In 4 vol./Bees Vol. 2 pp. 401–423./A D O

New ed./1823./
New ed. greatly enl./London, Tegg, 1837./In 3 vol./Bees Vol. 2. pp. 401–423./A D . M

Walker p. 23: 'Treats of bees in many countries.'

187 1821

The monarchy of the bees: a poem; illustrated with notes, exhibiting some of the most remarkable circumstances in the history of that little insect. / London, Goodhugh, 1821. / 2 leaves, 59 p. 15 cm. / A G O P R . M

2nd ed. / London, Simpkin & Marshall, 1822. / 2 leaves, 59 p. 15 cm. / B G

HACK, Maria 188 1821

Harry Beaufoy: or The pupil of nature. / London, 1821.

2nd ed. / London, Harvey & Darton, 1824. / vii, 194 p. / Bees pp. 137–154. / A

Hodgson 416.
Mamma tells Harry about bees including Hunter's experiments on the origin of beeswax.

COBBETT, William 189 1822

Cottage economy: containing information relative to the brewing of beer, making of bread, keeping cows, pigs, bees . . . etc. / London, C. Clements, 1822. / Bees pp. 170–174. / A

Many eds. incl. London, Peter Davies, 1926. / Bees pp. 122–125. / B

This short but well known passage recommends the cottager to improve his lot by keeping bees. Cobbett was an economist and politician, and lived for a time in the U.S.A.

MOUBRAY, Bonington *pseud.*
(LAWRENCE, John) 190 1822

A practical treatise on breeding . . . all kinds of domestic poultry . . . and on the management . . . of bees, by Bonington Moubray. 4th ed. / London, Sherwood, Neely & Jones, 1822. / 7 leaves, 312 p. illus. / Bees pp. 227–312. / A G O R S. M

5th ed. / 1824. / illus. / A G S
6th ed. enl. / 1830. / xii, 368 p. illus. / A G O
7th ed. / 1834. / x, 467 p. Bees pp. 280–332. illus. / S . M
8th ed. enl. / 1842. / A G O
Anr. ed. / 1853. / A
New ed. greatly enl. by L. A. Miall. / 1854. / A

The first three eds. have no bee material; in the 4th ed. this is 'from memoranda made during forty years practice'. The 7th ed. tells how a stock of bees was successfully transported to New South Wales in 1830–31.

LOUDON, John Claudius 191 1822
Encyclopedia of gardening./London, Longman Hurst, 1822./Apiary pp. 384–392, 723–741./A S
Subsequent eds./1825, 1828, 1835, 1850./A

RUSSELL, Robert 192 1822
Extracts on the natural history of bees. From F. Huber./Elgin (Morayshire), printed for and by Roel, 1822./17 p. together with *Extracts and observations on the history and management of bees.* From Key [*sic*], Bonner, &c./Elgin (Morayshire), printed by Roel, 1822./61 p./. C
2nd ed./*A typographical curiosity*: being a compilation on the natural history and management of bees./Elgin (Morayshire), 1834./vii, 136 p. 2 pl./A S
The author at the age of 15 cut the type and printed the book himself. Presumably the name Roel is derived from the first and last two letters of his name. The preface to the second edition gives an account of the printing and suggests that two copies of the first edition and twenty of the second were printed. Cornell University holds what is probably the only surving copy of the first edition, which was given by Russell to William Blackwood, an Edinburgh bookseller, to be presented to Sir Francis Freeling. A letter from Russell written at Emans Coffee House, 322 Strand, London, on 3 May 1834, which is bound into A's copy shows that Russell himself presented this copy of the second edition to Sir Francis Freeling whilst on a visit to London. The spine of A's copy of the second edition is inscribed *Rara liber.*

DICKSON, R. W. 193 [1823]
An improved system of management of live stock, and cattle, or A practical guide . . . to the nature, habits and proper management of all kinds of poultry, pigeons, bees and fish./London, Kelly, [1823]./Bees Vol. 1 pp. 458–479. illus./A
Illustrates an improved straw hive, very like the Greek one illustrated in Wheler's *A journey into Greece* 1682 (57), and also bee dress. Chapter on diseases.

EATON, Mary 194 1823
The cook and housekeeper's complete and universal dictionary . . . also original information relative to bees./Bungay (Suffolk), J. Childs, 1823./Bees pp. 26–29, 168./A S . W

MILTON, John 195 1823
The London apiarian guide for bee-keepers; comprising practical instructions to promote improvement in the cultivation of bees. With plates, and full directions for the use of hives in general, particularly the newly invented double topped straw hive. Published at the Apiarian Repository, 175 Strand where the finest honey, and hives of every description may be had./London, Author, 1823./21 p. illus./G S . C M

COCHRANE, James 196 1824
The modern receipt book—being a collection of nearly eight hundred valuable receipts, arranged under their respective heads; viz. domestic economy . . . management of bees, brewing, making wines . . . The whole carefully selected from the latest and best Authorities, and enriched by many original and valuable communications. Confirmed by actual experiment or personal observation./London, Newman, Dean & Munday, 1824./13 cm./Bees pp. 127–143./A O . S

SHAW, Thomas 197 1824
Recent poems, on rural and other miscellaneous subjects, by Thomas Shaw, apiarian, native of Saddleworth, Yorkshire./Huddersfield (Yorks.), Author, 1824./242 p. incl. front./A B

There is a long poem on bees and one on the seasons. In a note the author states: 'My manuscript treatise on the culture of the bee, may perhaps at some future period find its way to the press.' Unfortunately it seems that this manuscript was not published.

198 1825
Letters on entomology intended for the amusement and instruction of young persons./London, Whittaker, 1825./160 p. Bees pp. 58–75.

There is a copy in the Toronto Public Libraries.
Hodgson 411.

STRUTT, George 199 1825
The practical apiarian, or, A treatise on the improved management of bees. Clare (Suffolk), Shearcroft for the Author, 1825./1 leaf, viii, 112 p./ A G O S . M
Some copies have an inserted *Explanation of Strutt's model hive.*

[MORGAN, John Minter] 200 1826
The revolt of the bees./London, Longman Rees, 1826./2 leaves, 272 p./ A O S . M

Subsequent eds./1828, 1830, 1839, 1850, 1851, 1854./A B O S . M

Walker p. 49: '[The author was a] philanthropist and miscellaneous writer. Bees [were used here] as emblems of Society.'

BEVAN, Edward, M.D. 201 1827

The honey-bee; its natural history, physiology, and management./London, Baldwin, Cradock & Joy, 1827./xxvi, 1 leaf, 404 p. illus./A B D G H N O P R S . M W

Anr. ed./London, Van Voorst, 1838./xxiv, 447 p. illus. front./A B D G H R S . M W

Anr. ed. rev. enl. and illus. by W. A. Munn./London, Van Voorst, 1870. xxiv, 1 leaf, 384 p. 18 leaves, pl. I–IX, col. pl. A–L, front./A B G S . M W
This ed. is numbered 3rd ed. on the spine though not on the title page.

U.S.A. ed. 1843./Johansson 61.

The 2nd ed. was dedicated to Queen Victoria.
Walker p. 7: '[Bevan] practised as [a] Doctor at Mortlake, and at Congleton, Cheshire. His health giving way, he retired to Bridstow, near Ross [(Hereford.)], where he began to study bees. His book . . . at once . . . brought him into correspondence with the foremost bee-keepers of his time. In 1849 he moved to Hereford, where he died . . . Ob. [in] 'The Cottage Gardener', Lond., Vol. XXIII. Feb. 28, 1866.'
Bevan also wrote an article on the length of life of the different members of the bee community in *Magazine of Zoology and Botany* (1): 57–62 (1837).
The frontispiece (Fig. 9) shows the author's bee-house.

HOWATSON, T. M. 202 1827

The apiarian's manual; containing all that is important in the natural history of bees, or useful in their practical management./Edinburgh, Black & Tait, 1827./124 p. illus./G H P S . M

Walker p. 37: 'As a notable Scottish writer on bee-keeping, [he] worthily follows Maxwell, 1747 and Bonner, 1795. Though quite outclassed by Dr. Bevan's 'The Honey Bee', which appeared in the same year, this . . . treatise is a good practical work. In nat[ural] hist[ory] [this] follows Huber. [Howatson] designed an excellent Bar-hive.'

[JOHNSTONE, Mrs. Christian Isobel] 203 [1827]

Scenes of industry displayed in the bee-hive and the ant-hill./London, John Harris, [1827.]/xii, 212 p. 6 pl. incl. front. 17 cm./*History of bees* pp. 1–145./A B G N S . C

2nd ed. enl. with a brief description of the wonders of the insect world./

THE HONEY-BEE.

" What well appointed commonwealths ! where each
Adds to the stock of happiness for all ;
Wisdom's own forums ! where professors teach
Eloquent lessons in their vaulted hall !
Galleries of art ! and schools of industry !
Stores of rich fragrance ! Orchestras of song !
What marvellous seats of hidden alchymy !
How oft when wandering far and erring long,
Man might learn truth and virtue from the BEE !"—*Bowring.*

London, John Harris, 1830./xii, 280 p. 6 pl. incl. front. 18 cm./*History of bees* pp. 1–154./A B H S . M

For children./Hodgson 409.

A charming story which tells how Letitia Welbeck treads on a bees' nest and drops her book. A kind gentleman returns it to her and is able to reconcile her to the bees. Letitia and her school friends visit the gentleman, and by asking many intelligent questions they learn much about these industrious little insects. Each of the 6 pls. is subdivided into three, and all illustrate the bee story. Fig. 10 shows the plate opposite page 107 in the 1st edition.

GELIEU, Jonas de 204 1829

The bee preserver; or Practical directions for the management and preservation of hives: translated from the French of Jonas de Gelieu./ Edinburgh & London, J. Anderson, 1829./vi, 134 p. illus./A B G H P S . C M

2nd ed./Edinburgh, Edmonston & Douglas, 1876./xii, 84 p. illus./ A B G H S . M

For a review praising the book and the 'venerable translatress', see ***B.B.J.*** 3(35):206 (1876). The de Gelieus were a notable Swiss beekeeping family. The translation appeared two years after de Gelieu's death in 1827. The venerable translatress was Miss C. Stirling Graham.

205 1829

Natural history of insects./London, Murray, 1829./xix, 313 p. illus./In 2 vol. the second added 1835./xi, 352 p. illus./*The hive bee* Vol. 1 pp. 1–67 illus./A

2nd ed. 1830–38./Later ed. 1862. B G O S

Family Library Nos. 7, 51.

Walker cites Vol. 1 only and gives the date as 1827.

Pub. in U.S.A. 1830./Johansson 3./A has an ed. New York, Harper, 1843–46.

Fig. 9. Frontispiece of Edward Bevan's *The honey-bee* (1827, entry 201) showing his own bee-house, built of timber and thatched, and raised upon his brick potato-cellar.

The title page carries a quotation from Paley: 'A bee amongst the flowers in spring, is one of the cheerfullest objects that can be looked upon. Its life appears to be all enjoyment: so busy and so pleased.'

Fig. 10. One of the plates from *Scenes of industry* (1827, entry 203). On page 107: 'See, see!' cried Letitia, 'what a cloud of bees! . . .'. On page 108: 'I have heard', said Sarah, 'stories of brass pans and kettles being sounded when the bees swarmed . . .'. On page 112: Letitia, 'Look! look at the gardener, with a piece of gauze over his head, a great pair of gloves on, and a broom in his hand!'

LOUDON, John Claudius 206 1830
A manual of cottage gardening, husbandry, and architecture. / London, Spottiswoode for the Author, 1830. /illus. /Bees p. 8. Bee houses pp. 24, 28, 29, 38. / A

RHIND, William 207 1830
Studies in natural history; exhibiting a popular view of the most striking and interesting objects of the material world. / Edinburgh, Oliver & Boyd, 1830. / 1 engraving. /Of the economy of bees pp. 183–202, 219. / A S

[RENNIE, James] 208 1830
Insect architecture. / London, Knight, 1830. / xii, 420 p. *Architecture of the hive-bee* pp. 89–148. illus. / B

Subsequent eds. / 1830, 1845, 1846, 1857, 1869 enl. by Rev. J. G. Wood. / A B G O S . C M W

Vol. 1 of the Library of Entertaining Knowledge, published under the superintendence of the Society for the Diffusion of Useful Knowledge.

Rennie was Professor of Zoology at King's College, London.

[RENNIE, James] 209 1830
Insect transformations. / London, Knight, 1830. / xii, 420 p. illus. / A B S . M

Vol. 2 of the Library of Entertaining Knowledge (*see* 208).

Many references to bees.

[RENNIE, James] 210 1831
Insect miscellanies. / London, Knight, 1831. / xii, 414 p. / A B

Vol. 3 of the Library of Entertaining Knowledge (*see* 208).

Some reference to bees, and an index covering the three volumes.

CHILD, Lydia Maria (formerly FRANCIS) 211 1832
The girl's own book.
?5th ed. / London, Tegg, 1832. / 272 p. illus. /Bees pp. 215–223. Copy in Birmingham Institute of Education Library.

6th ed. / London, Tegg, 1833. /A (but copy destroyed by bombing in the Second World War).

13th ed. / London, Tegg, 1844. / 364 p. illus. /Bees pp. 324–331. / A
Numerous eds. to 1876; the later eds. revised by Mrs. R. Valentine.
'The hiving of bees is not a dangerous business for those who have experience in it; but children should never think of attempting it.'

Written for American children. Hodgson (403).

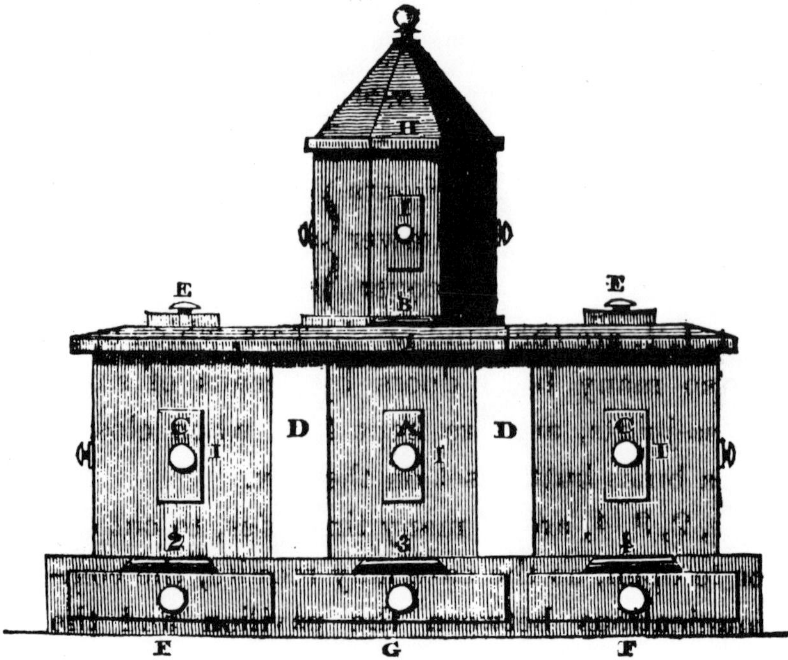
H
E
E
B
D
D
F
G
F

B
E
E
C
D
D
F
F
G

[FRANKUM, Richard] 212 1832
The bee and the wasp: a fable in verse, signed R.F./London, Tilt, 1832./23 leaves, illus. [by George Cruickshank]/A D

2nd ed./London, Basil Montagu Pickering, 1861./A D G S . M

Hodgson 404.

JESSE, Edward 213 1832
Gleanings in natural history with local recollections: to which are added maxims and hints for an angler./London, John Murray, 1832./xi, 313 p. 1 leaf./Bees pp. 14, 22, 167–174, 199./B

214 1832
Mamma's lessons for her little boys and girls. 3rd ed./London, Harris, 1832./Lesson XXI The bee pp. 109–113./A

We have not traced any earlier ed.
A very easy reader. For young children./Hodgson 405.

NUTT, Thomas 215 1832
Humanity to honeybees, or, Practical directions for the management of honeybees upon an improved and humane plan, by which the lives of bees may be preserved and abundance of honey of a superior quality may be obtained./Wisbech (Cambs.), Leach, 1832./xxiii, 1 leaf, 240 p. illus./A B D G H O R S . C M W

2nd ed./Wisbech (Cambs.), H. & J. Leach for the Author, 1834./xxii, 269 p. illus. 1 pl./B G N O S . C M W
3rd ed./Wisbech (Cambs.), H. & J. Leach for the Author, 1835./xxii, 269 p. illus. 1 pl. front./B G R S . C
4th ed. rev. enl. ed. by T. Clark./Wisbech (Cambs.), J. Leach for the Author, 1837./xxx, 281 p. illus. 1 pl. front./B G S . C M W

Fig. 11. Thomas Nutt's *Humanity to honey bees* (1832, entry 215) carries several engravings of his collateral bee-boxes. Here the upper illustration (from page 31) shows the hive in use, with central brood box, two side chambers and one upper one; below (from page 24) the hive is spread apart to show the communicating slots and the bell jar above full of comb honey. The middle drawer below was used for feeding, and the outer 'block-fronts' were movable and 'so contrived that ten thousand Bees can with ease leave their prison and their sweets in the possession of the humane apiarian, without the possible chance of a single intruder forcing its entrance to rob the magazine or to annoy the apiarian.'

5th ed. rev. enl. ed. by T. Clark./Wisbech (Cambs.), J. Leach for the Author, 1839./xxx, 281 p. illus. 1 pl. front./A B G P S . C M W
6th ed. rev. enl. ed. by T. Clark./Wisbech (Cambs.), J. Leach for the Author, 1845./xxxiv, 306 p. illus. 1 pl. front./A B G O R S . C M
7th ed. rev. enl. ed. by T. Clark as *Practical directions* . . ./Wisbech (Cambs.), J. Leach for the widow of the late T. Nutt, 1848./xxxiv, 306 p. illus. 1 pl. front./A B G S . M W

Nutt lived as a boy at Wisbech. A draper's assistant of no great schooling, he became interested in bees during a long convalescence. He asserts that he invented his collateral hive before ever reading a book on bees. He acquired many patrons and subscribers and made the hive and method very well known. One plate, which first appeared in the 2nd ed., is of the apiary at Delabere Park near Reading, the seat of one of Nutt's patrons the Marquess of Blandford who later became Duke of Marlborough. The sketch was made by the Duchess when Marchioness of Blandford, and Nutt complains in the 6th ed. that Cotton (238) had 'silently and surrepticiously' taken the sketch and placed it at the end of his list of bee books. This plate and the frontispiece are usually bound so as to pull out. There were three different German translations of the book and one Hungarian. Nutt was probaby among the first to use a thermometer in the hive. Collateral boxes had been previously described by S. White (1756) and W. Dyer (1781), and were later recommended by H. Taylor (1838) and J. Jones (1843); *see* 109, 137, 227, 246, respectively.
Nutt's hive is shown in Fig. 11.

[PAYNE, J. H.] 216 1832

The cottager's guide for the management of his bees upon the depriving system . . ./Bury St. Edmunds (Suffolk), Suffolk & Norfolk Apiarian Society, 1832./12 p. illus./M

For free distribution amongst the cottagers.

PAYNE, J. H. 217 1833

The apiarian's guide containing practical directions for the management of bees upon the depriving system./London, Simpkin & Marshall, 1833./xii, 71 p. 1 pl. (Nutt's hive). front. (cottage hive)/A B G S . M W

2nd ed./London, Simpkin & Marshall, 1838./viii, 100 p. illus. 1 pl. front./B S . C M

3rd ed. (styled 4th) *The beekeeper's guide.*/London, Newby, 1842./viii, 100 p. 1 pl. front./A B G R S . C M

4th ed. (styled 3rd) *The beekeeper's guide.*/London, Newby, 1846./xii, 108 p./G H N S . C M

5th ed. (styled 4th) *The beekeeper's guide.*/London, Newby, 1851./118 p./ A B G H . C M W

CHAMBERS, William 218 1833

Cottage economy./Edinburgh, W. Chambers, also London, Orr & Smith, 1833–35./Hints about bees p. 368./A S

Information for the People No. 46.

Subsequent eds./New & improved Series No. 81, 1842./Bees pp. 481–496./1848–49./ No. 42, 1856–57./The honey-bee pp. 657–672.

Issued periodically. The 1842 issues cost 1½d. each.

219 1833

Insect histories for children of eight or ten years old, by author of *Easy stories.*/London, Hailes, 1833./Bees pp. 111–168./A

Hodgson 402.

220 1833

Insects and their habitations: a book for children./London, Parker for the Society for Promoting Christian Knowledge, 1833./96 p. illus. 14 cm Bees pp. 6–34./A

Many subsequent eds. The 17th ed. is dated 1853. B has 1834 ed.

Hodgson 401.

221 1833

The wonderful history of the busy bees./London, Darton & Harvey, 1833./ 8 leaves each with hand col. pl./A

Hodgson 397 gives date 1839.

'When winter approaches, a terrible fate
Does the Queen and the bees of each city await;
For man, ever trying all ways to gain money,
Destroys the poor bees for their wax and their honey.'

BAGSTER, Samuel, Jun. 222 1834

The management of bees: with a description of the "Ladies' Safety Hive". /London, Bagster & Pickering, 1834./xx, 244 p. illus. col. front./A B G N O P S . C M W

2nd ed./London, Saunders & Otley, 1838./xvi, 240 p. 2 leaves, illus. col. front./B G H O R S . C M W

3rd ed./London, Saunders & Otley, [1852]./xvi, 240p. 2 leaves, illus. col. front./A B G H S . M W

Anr. 3rd ed./London, Griffin, 1865./xvi, 240 p./S . M

THE

MANAGEMENT OF BEES

WITH

A DESCRIPTION OF THE

"LADIES' SAFETY HIVE."

BY SAMUEL BAGSTER, JUN.

With Forty illustrative Wood Engravings.

LONDON:

SAMUEL BAGSTER, PATERNOSTER ROW; OR
VERE STREET, CAVENDISH SQUARE;
AND WILLIAM PICKERING, CHANCERY LANE.

MDCCCXXXIV.

Fig. 12. Title page of Samuel Bagster's book (1834, entry 222), printed in black and red. Opposite are splendid coloured drawings of queen, worker and drone, 'magnified' and also life size.

Walker p. 4: '[Bagster] followed his father as a publisher of Bibles and religious works. Memoir of his life by John Broad, Baptist Minister, 1837.' Biogr. *B.B.J.* 14(185): 3–5 (1886).
The title page of the 1st edition is shown in Fig. 12.

COOLING, Edwin 223 1837
The domestic gardener's assistant: in the modern practice of English gardening, designed for those who manage their own garden . . . to which are added remarks on the management of bees./Derby, Wilkins, 1837./Front. (skeps)./On the management of bees pp. 51–57./A

A CONSERVATIVE BEEKEEPER *pseud.*
(COTTON, William Charles) 224 [1837]
A short and simple letter to cottagers from a conservative beekeeper./Oxford, Oxford Apiarian Society, [1837]./24 p. illus./G N . C M
2nd ed./Hailsham (Sussex), Breads, [1838]./24 p. illus./A B G O P S . C
Anr. ed. . . . from a bee preserver./London, Society for Promoting Christian Knowledge, [1839]./24 p. illus./A B O . S
Walker p. 15 mentions 4 subsequent eds. The letter became part of *My bee book* (238).
There was a second letter to cottagers, apparently published three years after the first, but we have not seen a copy earlier than 1842, both letters then being included in *My bee book* (238). Walker cites the two letters together, dated 1843 and 1844.
Pub. in U.S.A. 1841./Johansson 110.

PATTERSON, Robert 225 1838
Letters on the natural history of insects mentioned in Shakespeare's plays./London, Orr, 1838./Letter 8 pp. 113–139./A D G O S . W
Anr. ed. *The natural history of insects mentioned* . . . 1841./C
Anr. ed./1842./A

PILE, Matthew 226 1838
The bee cultivator's assistant, or, A new and complete discovery in the management of bees; wherein is set forth a new and singular method of swarming . . . together with a full account of the nature and properties of the queens, and how they are invested with the power of raising a worker's egg to be a queen./Gateshead, (Durham), G. Watson, 1838./xi, 85 p. illus./B G S . C M (B's copy not illus.)

TAYLOR, Henry 227 1838

The bee-keeper's manual, or, Practical hints on the management and complete preservation of the honey-bee, and in particular in collateral hives. / London, Groombridge, 1838. / v, 1 leaf, 78 p. illus. / A G S . C M

2nd ed. enl. Title ends at *honey-bee.* / London, Groombridge, 1839. / viii, 126 p. illus. / D G H P R S . C M W

3rd ed. corr. / London, Groombridge, 1846. / vi, 144 p. illus. front. / A G O . W

Anr. 3rd ed. / London, Groombridge, 1849. / vi, 166 p. illus. With an appendix with an illustrated description of the improved 'amateur's bar-hive'. / A G S . C M W

4th ed. rev. and enl. / London, Groombridge, 1850. / viii, 184 p. illus. front. / B G . M W

5th ed. / London, Groombridge, 1855. / xiii, 216 p. illus. front. / A B G S . M W

Anr. ed. / 1859. / D

6th ed. . . . *honeybee; with a description of the most approved hives, and other appurtenances of the apiary*. With additions. / London, Groombridge, 1860. / xiv, 1 leaf, 224 p. illus. front. / A B G H P S . M

7th ed. *The bee-keeper's manual*; or The honey bee its management and preservation. With a description of the best approved hives, and other appliances of the apiary. Modernised and enl. by Alfred Watts. / London, Groombridge, 1880. / xii, 372 p. illus. front. / A B G H S . M W

Anr. 7th ed. / Edinburgh, Grant, [1880]. / xii, 372 p. illus. different front. / B

8th ed. Title as 7th ed. / London, Henry J. Drane, [n.d.]. / xii, 372 p. illus. front. / B

Watts included in the 7th ed. an appendix containing long extracts from Evans's poem *The bees* (170). He claims on the title-page to have been assistant editor of the 3rd ed. of Neighbour's *Apiary* (307) and says that he was responsible for adding several extracts to those given in the earlier editions.

WESTWOOD, John Obadiah 228 1838

The entomologists' text book: an introduction to the natural history of . . . insects. / London, Orr, 1838. / Hymenoptera pp. 371–389. illus. / A O S . C W

DOYLE, Martin *pseud.* **(HICKEY, Martin)** 229 1839

A cyclopaedia of practical husbandry and rural affairs in general. / Dublin, Curry, 1839. / Bees pp. 37–48. / A

New ed. enl. and rev. by Rev. W. Rham. / London, 1851.

SMITH, Richard 230 1839
The cottager's bee book: containing remarks on the conservative bee-keeper: twelve months' instruction on beekeeping; the condition of the bee considered, and how to improve it./Oxford, Author, 1839. 2 leaves, 104 p. illus./A O . C M

The conservative bee-keeper was Cotton who had published *A short and simple letter to cottagers* in 1837 (224).

WARD, M. A. 231 1839
The queen bee; or A fete to the blossoms./Bath (Somerset), Pocock also London, Simpkin Marshall also Hamilton Adams, 1839./iii, 93 p. 10 × 13 cm./G
A poem for children, telling how the queen bee because of her gratitude to the flowers invites them to a reception.

WESTWOOD, John Obadiah 232 1839
An introduction to the modern classification of insects./London, Van Voorst, 1839–40./Bees Vol. 1 pp. 233–262. Vol. 2 pp. 252–287. illus./ A D N O S . W

A BEE KEEPER 233 [c. 1840]
The cottage bee-hive; a letter to cottagers, showing a pleasant, profitable and humane method of keeping bees./Tunbridge Wells, (Kent), reprinted from *The cottager's monthly visitor*/16 p. illus./. M

2nd ed./London, Rivington, 1843./16 p. illus. front./B G . M

BOSWELL, Peter 234 1840
Bees, pigeons, and rabbits . . . the whole adapted as a text book for the young student./Glasgow, M'Phun, 1840./The bee pp. 1–24. 14 cm./ A O . M

Anr. ed./London, Routledge, 1852./Bees pp. 1–24, 14 cm./G

235 1840
Domestic animals: containing instructions for the rearing, breeding, feeding and management of . . . bees./London, Craddock, 1840./Bees pp. 57–60./A

Recommends keeping bees near London. When purchasing a swarm, 'send a hive ready to receive it'. For health—put a swarm in a new hive.

[DUNBAR, Rev. William] 236 1840
The natural history of bees, comprehending the uses and economical management of the British and foreign honey-bee; together with the

known wild species . . . with portrait and memoir of Huber./Edinburgh, Lizars, 1840./ix, 17–301 p. 30 pl. (not 36 as stated) mostly col., plus front. (portrait) and half-title./A B D G O P R S . M W

The Naturalists Library Vol. XXVI, Entomology Vol. VI.

Later eds. 1843, 1852, 1859, some not dated./B G O

The Naturalists Library was issued between 1833 and 1844. When the set was complete, a set of new title pages was issued, all dated 1843. The series was renumbered, and the volume on bees became Vol. XXXIV, Entomology Vol. VII. There is no change in the text of the various editions. Plate 12 and page 182 make it clear that Dunbar was the author, although the book is sometimes listed under the name of Sir William Jardine who edited the series and is shown (incorrectly) as the author on some later copies, or under the name of James Duncan who wrote the section on wild and foreign bees and is said in an advertisement to the 1840 edition to have 'conducted the volume through the press'. The coloured plates are particularly attractive. Dunbar also wrote three articles on bees in the *Edinburgh Philosophical Journal* 1820–24.

237 1841

The farmer's guide . . . for small farmers and cotter tenantry of Ireland./ Dublin, H.M. Stationery Office, 1841./183 p./Bees pp. 153–58./. M

ANCIENT LAWS AND INSTITUTES OF WALES 1841

Ancient laws and Institutes of Wales. (*See ms* 2).

COTTON, William Charles 238 1842

My bee book./London, Rivington, 1842./pagination very irregular—pages numbered to lvi, 2 leaves, then 60–368. (pp. 72–98 have the same no. on both left- and right-hand pages) illus. bibliog./A B G H N O P S . C M W

Reprint./Bath (Somerset), Kingsmead Reprints, 1977./lii, 1 leaf, 56–402 p. 1 leaf, illus. front. bibliog./B

In the reprint the erratum (p. 88) has been corrected (p. 101) and the pagination regularized.

This is in the nature of a scrapbook, containing Cotton's letters to cottagers, bee verses and mottoes, and reprints of parts of books by Southerne, Moore, Thorley, Sydserff, etc., and also the rules of the Oxford Apiarian Society, and details of Cotton's plans to introduce bees to New Zealand. Cotton, who was a clergyman, adopted a very moral tone in advocating the use of a puff ball narcotic to spare the lives of the bees. In 1848 he published *A manual for New Zealand beekeepers* in Wellington, and in 1872 translated Busch's poem *Schnurrdiburr* (324).

[JAMES, Thomas] 239 1842
Honey-bee and bee books. / *Quarterly Review* 71: 1–54 (1842). / A B S . M
Reprinted under title *The honey bee.* / London, Murray, 1852. / 2 leaves, 99 p. / A B G S . M W
Murray's Reading for the Rail containing works of sound information and innocent amusement: suited for all classes of readers.

JOHNSON, Cuthbert William 240 1842
The farmer's encyclopedia and dictionary of rural affairs . . . / London, Longman Brown, 1842. / Bees pp. 191–193. Honey pp. 636–637. Mead pp. 827–828. / A
An excellent and scholarly account mentioning Huish, Conservative Beekeeper, Bevan, etc.

LOUDON, Mrs. Jane *pseud.* **(WEBB, K.)** 241 1842
The year-book of natural history, for young persons. / London, John Murray, 1842. / Bees and wasps pp. 49–69. illus. / A S
Hodgson 394.

242 1842
Remarkable insects. / London, Religious Tract Society, 1842. / vi, 160 p. illus./Bees pp. 1–32. / A B S
Anr. ed. / London, Religious Tract Society, 1853. /
Hodgson 435. / The date (1799) given by Hodgson must be incorrect.

WIGHTON, John 243 1842
The history and management of bees, with notice of a newly-constructed hive by the Author . . . gardener to Lord Stafford. / London, Longman, also Norwich, Bacon, Kinnebrook & Bacon, 1842. 2 leaves, xii, 103 p. illus. /front. / A B G O S . C M W
Frontispiece shows Improved Polish hive.

[WILLIAMS, Rev. C.] 244 [1842]
The honey bee, the queen of the hive. / London, Religious Tract Society, [1842]. / 32 p. illus. / . M
2nd ed. / 1842. / H

BORRETT, W. 245 1843
The economist: a collection of choice receipts . . . for wines. / Diss (Norfolk), 1843. / Dyer's retrocoupling bee-boxes pp. 54–57.

JONES, John 246 1843

The eclectic hive: or, Directions for the management of Jones' Herefordshire collateral bee-boxes, including practical instructions, applicable to every system of bee cultivation./Hereford, Times Office, 1843./3 leaves, 54 p. illus./H S . M

A curious work with curious hive designs.

LLOYD, Rev. C. A. A. 247 1843

Lecture on the honey-bee delivered before the Natural History Society for Shropshire and N. Wales./Shrewsbury and London, 1843./35 p. illus./ . M

Walker p. 62 says this is 'unusually good'.

MILTON, John 248 1843

The practical bee-keeper; or, Concise and plain instructions for the management of bees and hives./London, Parker, 1843./viii, 147 p. 2 leaves, illus, front. bibliog./A B D G O S . M W

Contains many quotations from bee books, and a chronological list of writers.

New ed./London, Milton, 1851./63 p. illus. front./A B D G S . M W

Quotations and chronological list omitted.

New ed./*Bees and how to manage them*, or, The practical bee-keeper./London, Houston & Wright, 1865./63 p. illus./G S . W

Milton sold hives and honey at his Apiarian Repository, 175, Strand. He gives an interesting account of several foreign honeys. When Cotton went to New Zealand he appears to have left some notes with Milton, who used them in the 1843 edition of *The practical bee-keeper*. There was probably a row when Cotton returned, as the notes are omitted from the 1851 edition. Perhaps the row was partly caused by Milton's reference on p. 137 to 'the direful effects of the use of the fungus puff ball'. (*See* 238.)

MINTORN, John & Horatio 249 1844

The handbook for modelling wax flowers./London, Routledge, 1844./84p./A

We know of the following eds.

4th ed./London, Routledge, 1850./viii, 9–87 p./B

Dedicated to the Duchess of Northumberland.

7th ed./London, 1855./82 p./G

8th ed./1860./A

This book and Peachey's (272) are included as examples of the use of beeswax for this elaborate craft in the nineteenth century.

John Mintorn also published *Lessons in flower and fruit modelling in wax.*/London, [1870].

MUNN, William Augustus 250 1844
A description of the bar-and-frame hive invented by W. A. Munn, Esq. with an abstract of [Daniel] Wildman's *Complete guide for the management of bees throughout the year.*/London, Van Voorst, 1844./52 p. 1 pl./A D O
New ed./London, Van Voorst, 1851./52 p. 2 pl./A D G S . M

Munn also edited and revised the 3rd ed. of Bevan's *The honey bee* in 1870 (201). In the 1st ed. of his own book the frames pull out at the back, but in the 2nd ed. triangular frames pull up. Walker p. 50: 'Munn's hive, especially the form of it described in his first Ed[itio]n, 1844, was regarded by some bee-keepers as having anticipated the idea of Langstroth, and was used in litigation in the attempt to invalidate his patent.'

CHYLINSKI, Dobrogost 251 1845
The bee-keeper's manual: founded on the experience, during many centuries, of the apiarians in Poland./London, Orr, 1845./viii, 75 p. illus. front. 15 cm./A B G N O S . C M W

Anr. ed. *The bee-keeper.*/London, Houlston & Stoneman, [n.d.]/viii, 75 p. illus. front. 15 cm./Industrial Library/. M

The book recommends the use of a large Polish upright hive. The later ed. is the same as the 1st except for the title and has the same 3 pl. and front.

252 1845
The hand book of the honey bee, with hints on hives./London, Clark, 1845./58 p. illus. 12 cm. A O . M

Anr. ed./1850./58 p. illus./G S . M
Anr. ed./1852./G

Slaters Sixpenny Hand-books.

LOUDON, Mrs. Jane *pseud.* **(WEBB, K.)** 253 1845
The lady's country companion, or, How to enjoy a country life rationally./London, Longman Brown, 1845./Management of bees. pp. 343–348, 844./A S

SCUDAMORE, Edward, M.D. 254 1846
A description of bee-hives adapted for making artificial swarms; also, in good years, for taking some honey without destroying the bees./Canterbury (Kent), H. Chivers, 1846./23 p./. M

2nd ed. rev. and enl. *Artificial swarms*: a treatise on the production of early swarms of bees by artificial means . . ./London, Longman Brown, Green & Longmans, 1848./vi, 53 p./A B G O P S . M

[TOMLINSON, Charles & Sarah] 255 1846

Lessons derived from the animal world. The bee as an example of economy./London, Society for Promoting Christian Knowledge, 1846./Part 1 The hive bee 72 p. Part 2 The bumble bee. Part 3 The solitary bee. illus. 14 cm./G M

Second series./London, Society for Promoting Christian Knowledge, [1847]./xii, 291 p. illus. front./B O

Anr. ed./1851./illus./A

Hodgson 391.

AUBREY, John 1847

Natural history of Wiltshire . . . (*see* 63)

GOLDING, Robert 256 1847

The shilling bee book, containing the leading facts in the natural history of bees, with directions for bee management./London, Longman, 1847./58 p. illus./A B O S . C

2nd ed./London, Longman Brown, 1848./68 p. illus./A B D G P O S . M

Fraser p. 74: '[Golding] had helped in the preparation of [the 2nd ed. of] Bevan's *The honey-bee*, [201]. [Golding's book] contained a great deal of up-to-date information, and was a dangerous rival to Huish's book [176] which was much dearer.' Golding died at Hunton in Kent.

MAYHEW, Henry & Augustus 257 1847

The good genius that turned everything into gold: or The queen bee and the magic dress./London, 1847./A

Anr. ed./London, Routledge, [1879]./179 p. illus. by George Cruikshank./A

A Christmas fairy tale which tells how a woodman spares a nest of bees in a tree and meets a queen bee with magic powers.

[PATTISON, C. ?] 258 1847

Bees and bee-hives./Preface signed C.P./1847.

2nd ed./London, Biggs, also Tavistock (Devon), Brendon, 1848./31 p./. M

Based on a series of letters in a local periodical in 1846. We have not seen the 1st ed., but it is clear from the 2nd ed. that the book first appeared in 1847. The book is tentatively attributed to Pattison by Walker (p. 5), who added the name to his copy, now in M. The book advocates the system of Mr. Ceely of Eltham, near London.

RICHARDSON, H.D., F.R.P.S. 259 1847

The hive and the honey-bee: with plain directions for obtaining a considerable annual income from this branch of rural economy. To which is added, an account of the diseases of bees with their remedies; also, remarks as to their enemies, and the best mode of protecting the bees from their attacks./Dublin, McGlashan, 1847./107 p. illus./A B D G O . M

2nd ed./Dublin, McGlashan, 1849./108 p. illus./G H . M
New ed. much enlarged by J. O. Westwood./London, Orr [1852]./viii, 9–122 p. illus. front./A B G H P S . M
Anr. ed./Dublin & New York, Saxton, 1852./122 p./A . M W
Anr. ed./London, Warne, [n.d.]./122 p./B
Pub. in U.S.A. 1852./Johansson 454.

Westwood reduced the emphasis on collateral hives; he extracted, from his report on the insect products of the 1851 Great Exhibition, a description of the numerous beehives exhibited in the Crystal Palace.

FARMER'S LIBRARY 260 1847–48

Animal economy./London, C. Knight, 1847–48./Bees Vol. 2 pp. 605–624./A

Describes bee dress; quotes Sydserff, Cotton, Bevan (the best book), Pettigrew and Thorley.

261 1848

The bee-keeper's companion: comprising directions for the purchase of a stock of bees, account of the most superior hives in use, the food of bees, swarming, to avoid injury by stinging, method of inspecting the interior of a hive, securing the honey-comb, treatment of bees, etc., etc./London, E. Lloyd, 1848./18 p. illus./A . M

Lloyd's Illustrated Handbooks of Domestic Economy No. 2.
Contains illustrations of the Russian hive of Mr. Prokopovitch.

[COOPER, J. Fenimore] 261a 1848

The bee-hunter; or The oak openings. By the author of *The pioneers*, *Last of the Mohicans*, *Pathfinder*, *Deerslayer* etc. etc./London, Bentley, 1848./In 3 vol. vii, 314 p. 310 p. 300 p.

Anr. ed. *The oak openings*: or The bee hunter./London, Routledge, 1855./vii, 311 p. 1 pl. of bees and bears./A
Several eds., e.g. 1857, some not dated./A B G
Pub. in U.S.A. 1848./Johansson 105.

Fiction. Set in 1812 during the war between the U.S.A. and Canada. Contains an account of bee hunting.

HUNT, [James Henry] Leigh 262 1848.
A jar of honey from Mount Hybla./London, Smith Elder, 1848./200 p. illus./A

Anr. ed./London, Smith Elder, 1883./265 p. illus.

A mixture of old Sicilian story and English pastoral, with a chapter on bees and picture of Sicilian log hives.

MAUNDER, Samuel 263 1848
The treasury of natural history, or, A popular dictionary of animated nature./London, Longman Brown, 1848./Bees pp. 57–60. illus./S

2nd ed./London, Longman Brown, 1849./A

Other eds. 1854, 1862, 1870 rev. and corr. by T. Spencer Cobbold, 1873 rev. and corr. by E. W. H. Holdsworth, 1874.

ACHETA DOMESTICA *pseud.*
(BUDGEN, L. M.) 264 1849
Episodes of insect life by Acheta domestica./London, Reeve, 1849–51./ In 3 vol./Bees Vol. 1 pp. 232–263. illus./A S
Anr. ed. rev. by J. G. Wood,./London, Bell & Daldy, 1867./xv, 430 p. illus./A

Anr. ed./1879./xv, 430 p.

Acheta domestica is the house cricket or 'cricket on the hearth'.

[ELLIS, Robert] 265 1849–50
The life of an insect./London, Society for Promoting Christian Knowledge, 1849–1850. In 2 vol. Vol. I 1849. Vol. II, 1850./viii, 384p. XII pl./

There are frequent references to bees in the text and plates. Volume II has a plate illustrating (somewhat inadequately) part of Huber's researches on the bee's use of the antennae.

Robert Ellis, F.L.S., was surgeon to the National Society's Training Institution and scientific editor of the Catalogue of the Great Exhibition of 1851. He also wrote *The life of a bird* and *The life of a tree.*

MINER, T. B. 266 1849
The American beekeeper's manual: being a practical treatise on the history and domestic economy of the honey-bee, embracing a full illustration of the whole subject, with the most approved methods of managing this insect through every branch of its culture, the result of many years'

experience. / London, Wiley, 1849. / iv, 5–349 p. illus. front. / A B D G H O P S . M

Subsequent eds. / 1851, 1852, 1857, 1860. / B G P. W

Pub. in U.S.A. 1849. / Johansson 344.

Walker p. 48: '[The] best all-round American bee-book till then published.'

OWEN, Richard 267 1849

On parthenogenesis, or, The successive production of procreative individuals from a single ovum. / London, Van Voorst, 1849. / 76p. illus. / A D G O . M

VIRGIL (Publius Vergilius Maro) 268 1849

The works of Virgil; translated by Charles Rann Kennedy. / London, Stevens, 1849. / Fourth Georgic Vol. 2 pp. 146–170. / A

Anr. ed. / London, H. G. Bohn, 1861, / 1 leaf, 453 p. Fourth Georgic pp. 104–122. / B

Verse.

1850 to 1899

The change to rational beekeeping

BEVAN, Edward 269 1851
Hints on the history and management of the honey-bee./Hereford, Times Office, 1851./23 p./A P.M

The substance of two lectures read before the members of the Hereford Literary, Philosophical and Antiquarian Institution.

DOYLE, Martin *pseud.* **(HICKEY, M.)** 270 [1851]
Rural economy for cottage farmers and gardeners: a treasury of information on cow keeping . . . honeybees, etc./London, Groombridge, [1851]./The honey bee pp. 106–146. illus./A. M

New ed./London, 1860./Honey bee pp. 77–106. illus./G. M

Economic Library Vol. 1.

A COUNTRY CURATE *pseud.*
(FILLEUL, P. V. M.) 271 1851
The English bee-keeper, or, Suggestions for the practical management of amateur and cottage apiaries, on scientific principles . . . by a country curate./London, Rivington, 1851./xvi, 212 p. illus./A B G S . M W

Pub. in U.S.A. 1851./Johansson 169.

Describes a window apiary.

PEACHEY, Mrs. Emma 272 1851
The royal guide to wax flower modelling by Mrs. Peachey, artiste to Her Majesty./London, Author, 1851./xvi, 72 p. 4 col. pl. incl. front./A B

See also Mintorn (249).

BEESLEY, James 273 1852
A practical monthly bee calendar./London, 1852./7 p./A N O S . M

See W. White (123).

Beesley belonged to an old Quaker family of Banbury. His calendar is included in the 1852 ed. of White's book, and no separate publication of it has been traced.

[CROSS, J. H.] 274 [1852]
The hive and its wonders./London & Edinburgh, Gall & Inglis, [1852]./ 124 p. illus. 13 cm./A B G S . M

New and rev. ed./London, Religious Tract Society, [1853]./126 p. illus. front. 16 cm./A B G S . M
Anr. ed./London, R.T.S., [1876]./126 p. illus./A S
Rev. ed./London, R.T.S., [1881]./126 p. illus./A

Hodgson 389.

Written for the American Sunday-school Union, 1851./Johansson 111.

(*cancelled*) 275

DESBOROUGH, J. G. 276 1853
On the duration of life in the queen, drone and worker of the honeybee. To which are added observations on the importance of this knowledge in deciding whether to preserve stocks or swarms. Being the Prize Essay of the Entomological Society of London for 1852./London, printed for Entomological Society by C. Roworth, 1853./31 p./A S. M

Walker p. 17 also has 'Observations on the Honey Bee, in continuation of Prize Essay. Read April, 1855. *Ex Trans. Ent. Soc. Ser. II.*, Vol. III., pp. 187–196.'
Desborough's library went to the British Bee-keepers' Association.

PRATT, Mrs. John Burnett 277 1853
The bee and bee-keeper's friend./Aberdeen, Smith, 1853./16 p. illus./.M

Advocates use of puffball in fumigation.

WOOD, Rev. J[ohn] G[eorge] 278 1853
Bees: their habits, management and treatment./London, Routledge, 1853./ 1 leaf, 114 p. illus. front. 18 cm./A G S . M W

2nd ed./London, Routledge, 1855. Described by Walker as 'a page-for-page reprint; fresh front.' 16 cm./. M
This ed. is described as 2nd ed. on front cover but not on title page.

New ed./London, Routledge, 1858./1 leaf, 114 p. illus. front. 16 cm./ B S. M

New ed./London, Routledge, 1863./1 leaf, 114 p. illus. front. 16 cm./. M

New ed./London, Routledge [n.d.]/2 leaves, 114 p. illus. front. 16 cm./ B G S. M
This ed. is bound in orange publisher's cloth. We have seen various copies with different advertisements, including one for Neighbour's *The apiary*

2nd ed. (1866), and copy in M has reference to *Every boy's annual for 1869*. Similar copies [n.d.] paper-covered boards./B S

New ed. London, Routledge, [n.d.]/2 leaves, 114 p. illus. no front. 16 cm. grey cloth./B S B's copy contains advertisement dated 1884.

Anr. ed. London, Routledge, [n.d.]/2 leaves, 114 p. illus. no front. 18 cm. paper covered boards./B has two copies which contain Routledge's *Railway Library Advertiser* for 1893 and 1895.

Books for the Country No. 11.

Pub. in U.S.A. 1863./Johansson 608

Although this book appeared in many different forms with different frontispieces or none, the text is the same in each ed. page for page. Rev. John Wood was a popular lecturer on many subjects and wrote many books on natural history.

GATTY, Mrs. Alfred 279 1855

Parables from nature./London, Bell & Daldy, 1855./xiv, 144 p. illus. by the author, 12 cm./A

Originally there were eight parables, of which *The law of authority and obedience* (pp. 17–34) extols virtues found in the beehive. A 2nd ed. appeared in 1855, and there were 18 eds. of the first series up to 1882. There were also four further series of parables and numerous collected editions, notably London, Bell, 1880, with illustrations by Holman Hunt, Burne Jones, Weir, Tenniel and others; also London, Bell, 1910, with illustrations by Alice B. Woodward. B has photocopy of the bee parable from this ed.

KARR, Alphonse 280 1855

A tour round my garden. Translated from the French of Alphonse Karr. Revised and edited by the Rev. J. G. Wood, M.A., F.L.S. &c. With one hundred and seventeen illustrations by William Harvey./London, Routledge, 1855./xii, 332 p. illus./A

Bees on pp. 184–198, also references to the mason bee and leafcutter bee.

PAYNE, J. H. 281 [1855]

Bee-keeping for the many: being plain instructions for the . . . management of the honey bee./London, Cottage Gardener Office, [1855]./55 p. illus./.M

Anr. ed./London, Cottage Gardener Office, [1856]./54 p. illus./A O S

5th ed. rev. and enl. as *Beekeeping for the many*; or, The management of the common and Ligurian honey bee, including the selection of hives and a bee-keeper's calendar. / London, Journal of Horticulture Office, [1863 ?]. 70 p. illus. / G. M

?7th ed. title as above / 70 p. illus. / A B G S . M

Manuals for the Many No. 3.

We have compared the two copies at G. The first states on the title page that it is the 5th ed. The date 1863 ? has been written on. The publisher's address is 162 Fleet St. The second is undated and described as 'Thirty sixth thousand'. The publisher's address is 171 Fleet St. Text and illustrations are very similar but not identical.

PLINY (the elder, **Caius Plinius Secundus**) 282 1855
Natural history of Pliny; translated with copious notes and illustrations by John Bostock and T. H. Riley. / London, Bohn, 1855–57. / 6 vol. / Bees Vol. 3 pp. 5–24. Vol. 4 pp. 339–346. / A G

Bohn's Classical Library.

SMITH, Frederick 283 1855
Catalogue of British Hymenoptera in the British Museum. / London, British Museum, 1855. / Apidae—Bees Part 1. illus. / A G S . M

2nd ed. *Catalogue of the British bees in the collection of the British Museum.* / 1876. / xii, 236 p. illus. X pl. + 1 pl. / A B S. M
New ed. / 1891. / illus. / A S

A.L.O.E. (A lady of England) *pseud.*
(TUCKER, Charlotte Maria) 284 1855
Wings and stings: a tale for the young. / London & Edinburgh, Nelson, 1855. / 99 p. illus. / A

Other eds. / 1858, 1864, 1870, 1872, 1885, 1888. / A R S; B has 1885, 1888, both 108 p. 9 pl. front.

Hodgson 387.
Pub. in U.S.A. 1872. / Johansson 574.

HIBBERD, Shirley 285 1856
Rustic adornments for homes of taste, and recreations for townfolk in the study and imitation of nature. / London, Driffield, 1856. / Apiary pp. 255–326. illus. / A

2nd ed. / London, 1858. / Apiary pp. 255–326. illus. / S

New ed. rev. / London, Groombridge, 1870. / Apiary pp. 229–276. illus. / A S
New ed. / London, Collingridge, 1895. / A

Contains much information on the life of the bee, and on various feeders and fumigators. Interesting changes between editions in type of hive recommended.

286 1856

The inquisitive bee, and other tales./London, Cash, 1856./96 p./A

The inquisitive bee observes the romance between Alfred and the governess Miss Atley, and then tells other tales.

For children.

LARDNER, Dionysius, D.C.L., *editor* 287 1854–56

The museum of science and art. A periodical issued in penny numbers which when completed consisted of 12 single or 6 double volumes. Various parts of the work were also issued separately, e.g.:

The bee and white ants, their manners and habits; with illustrations of animal instinct and intelligence./London, Walton & Maberly, 1856./5 leaves incl. front./*The bee* pp. 1–112 illus., *The white ants* pp. 97–128, *Instinct and intelligence* pp. 113–176 (Bees are mentioned pp. 140–144)./B

Bee portion only as *The bee. Its character and manners.* Later editions published by Lockwood [n.d.]

A B G S. M W B has Vol. X of the single volumes.

S., S. S. 288 1856

The hive and its commonwealth./London, Ford, 1856./69 p./A B

A scholarly account quoting Réaumur, Cotton, Bevan, Hunter and Huish.

BEST, Henry 1857

Rural economy in Yorkshire in 1641. (*see* 28)

SIEBOLD, Carl Theodor Ernst von 289 1857

On a true parthenogenesis in moths and bees: a contribution to the history of reproduction in animals. Translated by William S. Dallas./London, Van Voorst, 1857./viii, 110 p. 1 pl./Honey bee pp. 39–99./A B G S . M W

MILLER, Thomas 290 [1858]

Birds, bees and blossoms; original poems for children./London, Brown, [1858]./illus. by Birket Foster./Bees pp. 41–47./A S. M

Anr. ed./Edinburgh, [1864]./A

Hodgson 385.

DARWIN, Charles 291 1859
On the origin of species by means of natural selection, or the preservation of favoured races in the struggle for life./London, Murray, 1859./ix, 502 p. 1 folding diagram. Many editions in Britain and U.S.A. Many translations. For details see Freeman, *The works of Charles Darwin: an annotated bibliographical handlist*, 1965, 2nd ed. rev. and enl. 1977.

Described by Freeman as the most important biological book ever written, Darwin's *Origin of species* includes descriptions of experiments he carried out on comb building by bees. There are also important references to bees in *On the various contrivances by which . . . orchids are fertilised by insects . . .* (1862), *The effects of cross and self fertilisation in the vegetable kingdom* (1876), and a paper in the *Gardeners' Chronicle* in 1858: *On the agency of bees in the fertilisation of papilionaceous flowers and on the crossing of kidney beans*.

See the article *Charles Darwin and bees* by A. D. Brian and E. Crane, *Bee World* 40(12): 297(1959). The IBRA Library holds the original of a letter written by Darwin to T. W. Woodbury (*see* 302) thanking him for sending some comb foundation 'which interested me much'.

[LANCASTER, Edwin] 292 [1859]
Half hours with the microscope. Being a popular guide to the use of the microscope as a means of amusement and instruction. By E. L./London, Hardwicke, [1859]./Illus. by Tuffen West./Head and mouth of bee, fig. 194 p. 64; foot of bee, fig. 199 p. 67; crystals of honey, fig. 228 p. 24; crystals of sugar from adulterated honey, fig. 230 p. 24./A

Many eds.

HERMANN, H. C. 293 1860
The Italian alp-bee, or, The gold mine of husbandry. Short and practical instructions to breed genuine prolific Italian queens; to multiply them by hundreds in a few months and how to change German hives into Italian. London, G. Neighbour, 1860./32 p./A G S. M

'A young, yellow, impregnated queen from March 15th to April 30th with 500 bees for 20 shillings: with 1000 bees for 35 shillings. Italian drones till July 30th 2 shillings per 100. Italian Queen cells 3/– each. H. C. Hermann Bee Cultivator.'
This was the first introduction of Italian bees to England. Neighbour sold the queens, and Woodbury used and recommended them.

LANCASTER, Edwin 294 1860
The uses of animals as applied to the industry of man./London, 1860./ The bee 10 p./A

Anr. ed./1876./A S

MACKENZIE, John 295 1860
The management of bees./Edinburgh & London, Blackwood, 1860./32 p./A S. M

SAMUELSON, James 296 1860
The honey-bee; its natural history, habits, anatomy and microscopical beauties. Also two chapters on instinct and reason; being an introduction to the study of comparative psychology by the same author [J. Samuelson] assisted by J. Braxton Hicks./London, Van Voorst, 1860./xvi, 166 p. VIII pl. incl. front./A B G R S . M W

Humble Creatures Part II; Part I dealt with the earthworm and the common housefly.

TEGETMEIER, William Bernhard 297 [c. 1860]
Bees, hives and honey: handy helps to useful knowledge./London, W. Allan, [c. 1860]./20 p. illus./G. M

Anr. ed./London, W. Allan, 1865./39 p. illus./G S . M

Tegetmeier was a founder of the Royal Entomological Society. He wrote to *The Field* till he was over 90. Walker (p. 16) had a book containing 'The correspondence (somewhat heated) that appeared in *The Times*, *Standard*, *Field*, *Daily News*, *Morning Post*, and other journals on the articles of The Times Bee-master [Dr. John Cumming: *see* 304]. Collected and mounted in book-form . . . by Tegetmeier, his chief opponent.' Walker p. 68: 'Obit. notices in *The Field*, Nov. 28th, 1912; *B.B.J.*, Dec. 12th, 1912'.

HUNTER, John, F.R.S. 298 1861
Essays and observations on natural history, anatomy, physiology and geology, by John Hunter, F.R.S. being his posthumous papers on these subjects, arranged and revised, with notes . . . by Richard Owen, F.R.S., D.C.L./London, Van Voorst, 1861./The hive-bee [*Apis mellifica*, Linn.] is in Vol. 2 pp. 462–466./A(B)G

See Hunter 146

BEETON, Samuel Orchart 299 [1861–62]
The book of home pets./London, S. O. Beeton, [1861–62]./xvi, 832 p. illus. Issued periodically in threepenny parts./Bees pp. 705–729 in Part XXIII entitled *White and other fancy mice, silkworms, etc.*, although the part contains only the bee and the silkworm./A

Pages 353–832 reissued in 1870 as *Beeton's book of poultry and domestic animals.*/London, Ward Lock, [1870]./xii, pp. 353–832./A

It appears from advertisements that Ward Lock also issued the whole work under the title *Beeton's book of home pets.*

Part XXIII also appeared under the more apt title *Bees, silkworms, etc.*/ London, Bickers & Bush, [n.d.]./pp. 705–736./A G . M

And as *Bees, silkworms, and inhabitants of the aquarium*, also a chapter on ferns./London, Bickers & Bush, [n.d.]./pp. 705–736, 769–832./A G S . M W

BEETON, Samuel Orchart 300 [1861–62]

The book of garden management: comprising information on laying-out and planting gardens . . . monthly operations in each department . . . the management of bees./London, S. O. Beeton, [1861–62]./viii, 760 p. illus./ Bee-keeping Chapter XXV pp. 433–454. illus./A

Pub. in monthly parts entitled *Beeton's book of garden management and rural economy.*

The book claims 'no more merit than belongs to a fairly-balanced compilation, drawn from widely spread but sound sources of information'. Bees are also mentioned in the monthly calendar for August, September, October. In later editions, published by Ward Lock, coloured plates of flowers appeared, and the bees were omitted. The companion work *Beeton's book of household management*, edited by Beeton's wife Isabella, achieved a more lasting fame.

RICHARDSON, Mrs. 301 [1862]

The story of a bee and her friends—told by herself./London, Wertheim, Macintosh & Hunt, [1862]./1 leaf, vi, 1 leaf, 63 p. 6 pl. front./A B

Anr. ed./London, Macintosh, 1867./1 leaf, vi, 1 leaf, 63 p. 6 pl. front./ B G S . M

For children./Hodgson 379.
Walker p. 60: 'very well got-up'.

WOODBURY, T. W. 302 1862

Bees and bee-keeping; with an account of the introduction of the Ligurian species of honey-bee into Great Britain./London, Clowes, 1862./1 leaf, 18 p. illus./(B)S . M

Reprint from *Bath and West of England Agricultural Journal* 10(1): 84–101 (1862).

Woodbury introduced to Britain the first hive embodying Langstroth's incorporation of the bee space. B has the copy of Langstroth's book *The hive and the honey-bee* (Johansson 284) that he presented to Woodbury. Woodbury recommended Italian queens. Neighbour (307) records that Woodbury wrote articles under the name A Devonshire Beekeeper. An article thus signed in *The Cottage Gardener and Country Gentleman*, Dec. 7, 1859, records how Cotton's plan (238) to take bees to New Zealand failed when the sailors threw the bees overboard in bad weather.

NECKHAM, Alexander 1863
De naturis rerum. (*See ms* 5)

PETTIT, Wyatt J. 303 [1863]
The management of bees: with an illustrated catalogue of hives and apiarian furniture. / Dover (Kent), Brett, [1863]. / 24 p. illus. / A G . M

2nd and enl. ed. / Dover (Kent), Brett, [1867]. / iv, 48 p. 2 leaves, illus. front. / A (B) G P S . M

G has the 2nd and enl. ed. as above bound in boards with publisher's cloth price 1/6d, and also a copy in paper 48 p. without the half title and 2 leaves of index, price 1/–, on which the date 1866 has been written. Also Pettit's *Illustrated catalogue* abridged ed. 16 p. illus. [n.d.] and a later *Illustrated catalogue* 16 p. illus. on which the date 1881 has been written.

Pettit made and sold beekeeping equipment in Dover. His equipment was recommended by The Times Bee-master (*see* 304).

'THE TIMES' BEE-MASTER *pseud.*
(CUMMING, John) 304 1864
Bee-keeping. By 'The Times' Bee-master. / London, Sampson Low, Son & Marston, 1864. / xx, 224 p. illus. / A B G R S . M W

2nd ed. / London, Sampson Low, 1871. / xx, 224 p. illus. / A G S . M

The author had previously conducted a heated correspondence with Tegetmeier and others in *The Times*, *Standard*, *Field*, etc.; see Tegetmeier (297). The book owes its origin to a number of long letters written to *The Times* and reflects some of the controversy. The 2nd ed. appears to be a reprint with a new preface. Cumming was a minister of the Scottish National Church.

PHIPSON, Thomas Lamb 305 1864
The utilisation of minute life: being practical studies on insects . . . / London, Groombridge, 1864. / Insects producing wax, resin, honey and manna pp. 65–90. / A S

WOOD, Rev. J[ohn] G[eorge] 306 1864
Our garden friends and foes. / London, Routledge, 1864. / Bees pp. 261–271. / illus. / A S

Reissued 1882. / A

NEIGHBOUR, Alfred 307 1865
The apiary; or, Bees, bee-hives and bee-culture: being a familiar account of the habits of bees, and the most improved methods of management, with full directions, adapted for the cottager, farmer or scientific apiarian. /

EXTERIOR OF AN APIARY.
As originally erected in the Zoological Gardens, Regents Park.

INTERIOR OF THE ABOVE.
May be taken as suggestive for the construction or appropriation of rooms for larger Apiaries in summer houses or other outbuildings.

Fig. 13. Alfred Neighbour's bee house, from *The apiary; or, Bees, bee-hives, and bee culture* 1865 (entry 307).

London, Kent & Neighbour, 1865./xi, 134 p. 1 leaf, illus. (but no pl.) 23 cm./A G S . M W

2nd ed./London, Kent & Neighbour, 1866./2 leaves, i–viii, xv–xviii, 271 p. 269–274, iii. illus. 1 pl. front. 18 cm./A B G S . M
3rd ed. rev. and enl./London, Kent & Neighbour, 1878./1 leaf, xxvi, 359 p. 2 leaves, illus. 1 pl. front. 18 cm./A B G P S . M

Geo. Neighbour and Sons were considerable appliance dealers. The book contains their catalogue, and an account of their stand at the 1851 Exhibition and at the International Exhibition of 1862 (with a picture). There is also the text of an advertisement: 'June 20, 1772. Exhibition of bees on horse back . . . The celebrated Daniel Wildman will exhibit several new and amazing experiments, never attempted by any man in this or any other Kingdom before.' Alfred Watts claimed to be the assistant editor of the 3rd ed. See note under Taylor (227).
The outside and inside of Neighbour's bee house are shown in Fig. 13.

Ancient laws of Ireland 1865–1901
See ms 1.

COX, Irvin Edward Bambridge *editor* 308 1866
The farm, garden, stable and aviary: to which is added some useful information relating to the poultry-yard, beekeeping and natural history; being alike valuable to country gentlemen, farmers, gardeners, and those interested in the preservation, etc. of birds, beasts and fishes. Edited by I.E.B.C./London, Cox, 1866./Appendix, The beekeeper and naturalist pp. 119–139./illus./A S

2nd ed./1869–71./illus./A

Field Library Vol. III.

ROLLE, Richard 1866
English writings. (*See ms* 10)

SHUCKARD, W[illiam] E[dward] 309 1866
British bees: an introduction to the study of the natural history and economy of the bees indigenous to the British Isles./London, Reeve, 1866./xvi, 371 p. XVI col. pls. incl. front./*Apis mellifera* pp. 321–362./ A B G S . M W

Later issue [n.d.]/B
Later issue [n.d.]/pls. not col./B
Later issue Ashford (Kent), Reeve, [n.d.]

There were clearly several issues. B has four copies. The 1st dated 1866 contains Reeves' catalogue which is undated; the 2nd dated 1866 contains

Reeves' catalogue dated 1870; the 3rd, undated, contains Reeves' catalogue undated; the 4th in green cloth is undated, with no catalogue and the plates not coloured. D.A.S. has a similar copy with Reeves' London address covered by a label giving an address in Ashford, Kent. In the earlier issues the plates are coloured by hand and are particularly fine. As might be expected, the colouring of the plates in the first three B copies differs slightly.

WINTLE, L. & M. 310 1866

Tot and the cat . . . [*and*] *Mrs. Bee.*/London, Bell, 1866./*Mrs. Bee* pp. 81–108.

Subsequent eds./1884, 1920./A S

Anr. ed./London, Bell, 1927./1 leaf, 108 p. 2 leaves, illus. 3 pl./*Mrs. Bee* pp. 81–108./B

Bell's Books for Young Readers./Hodgson 378.

FILLEUL, P. V. M., M.A. 311 [1867]

Profitable bee-keeping on improved principles: chiefly designed for the use of cottagers./London, Society for Promoting Christian Knowledge, [1867]./79 p. illus. incl. front. 14 cm./. W

Anr. ed./London, S.P.C.K., 1868./79 p. illus. incl. front. 14 cm./ A B G S . M

Rev. ed. with additional matter./London, S.P.C.K., 1883./92 p. illus./ A G . M

Reprint from *People's Magazine* for 1867.

FIGUIER, Louis 312 1868

The insect world, being a popular account of the habits and economy of some of the most interesting species. Edited from the French by Y.D./ London, 1868./Bees pp. 312–360./A

Anr. ed. translated and revised by E. W. Janson./1869./Bees pp. 312–402./S

New ed. rev. and corr. by P. M. Duncan./London, Cassell, Potter & Galpin, 1872./Bees pp. 313–357. illus./A G

Anr. ed. rev. and corr. by P. M. Duncan./London, Cassell, 1892./A

This book contains interesting illustrations of various European hives and equipment. The original work was *Les insectes*, Paris, 1867./A

PAGDEN, J. W. 313 [1868]

£70 a year: how I make it by my bees; and how a cottager or other may soon do the same./Alfriston (Sussex), Author, [1868]./35 p. 14 cm./. M

There were at least 18 editions, none of them dated, and it seems likely that they are really reprints of the 1st ed. Because of its small size, not many copies have survived./B (18th ed.) G S

Pub. in U.S.A. *$350 a year* . . . [1868?]./Johansson 396.

SHERER, John 314 [1868]

Rural life described and illustrated. . . . London, London Printing and Pub. Co., [1868]./27 cm./The bee and the management of the apiary pp. 994–1002./A S . M

GOULDING, W. & H. M. 315 1869

Bees. Presented by W. & H. M. Goulding, manure manufacturers./ Carlisle, 1869./47 p. illus.

The book was printed by Hudson Scott in Carlisle, although Gouldings 'Manure manufacturers by royal letters patent' had their works in Cork and Dublin.

WOOD, Rev. John George 316 1869

Bible animals./London, Longmans Green, 1869./Bees pp. 605–613./A

317 [1870]

Mrs. Martin's bee hive./Oxford & London, [1870]./28 p./A . M

Tracts for Cottagers No. 3.

PETTIGREW, A. 318 1870

The handy book of bees: being a practical treatise on their profitable management./Edinburgh & London, Blackwood, 1870./xii, 1 leaf, 193 p. illus./A B G S . M W

2nd ed. rev. and improved./Edinburgh & London, Blackwood, 1875./ xiv, 162 p. illus./A B G S . M W

3rd ed. rev./Edinburgh & London, Blackwood, 1875./xiv, 162 p. illus./ B S

4th ed. rev. and enl./Edinburgh & London, Blackwood, 1881./xv, 172 p. illus./A B G P S

5th ed./Edinburgh & London, Blackwood, 1889./xv, 172 p. illus./S

All eds. from 2nd appear to be the same, except that 4th and 5th eds. have a 10-page supplement at pp. 145–154. All copies that we have seen have similar publisher's green cloth bindings and contain various versions of Blackwood's book catalogue.

Pettigrew advocated a much larger skep than was usual, and obtained more honey from it.

VIRGIL (Publius Vergilius Maro) 319 1870

The story of Aristaeus and his bees: a translation of Virgil's Fourth Georgic into English heroics by R. M. Millington./London, Longmans, 1870./ 53 p./A

New ed. rewritten./London, Longmans Green, 1870./88 p./A G
Verse.

EDWARDS, Kinard Baghot
(afterwards Baghot de la Bere) 320 [1871]

The French Bishop's advice to his poor clergy. 'Keep bees! Keep bees!': the A B C guide or cottager's manual, shewing how, by proper management, ten stock hives will return the owner an annual profit of £50 sterling. Part 1 London, Hinckley, [1871]. Part 2 Keep bees, Keep bees. Containing illustrated designs for . . . straw and box hives, patent swarm signal and German drone-catcher, American nest hive and patent bee-feeder . . . concluding with . . . the driving, shaking, chloroforming and fumigating of bees./London, Hinckley, [1871]./In 2 parts, 16 p. & 8 p. illus./ A G S . M

Anr. ed. *The French Bishop's advice* . . . 10 or 12 stock hives . . . profit of £40 or £50 sterling. Part 2 Patterns of the best and most approved form of bee hives. Honey extractor, drone and bee traps, feeders &c, &c., concluding with remarks on the uniting of swarms, and the chloroforming, 'driving', and 'shaking' of bees./Hinckley (Leics.), [Author, 1876]/In 2 parts, 22 p. & 25–36 p. illus./A B S . M

Walker p. 21 gives date of 2nd ed. as 1876.

STAVELEY, E. F. 321 1871

British insects: a familiar description of the form, structure, habits and transformations of insects./London, Reeve, 1871./Bees pp. 221–250./ illus./A S

[WOOD, T.] 322 [1871]

Curiosities of entomology./London, Groombridge, [1871]./64 p. 10 col. pl./ Bees pp. 5–17./illus./A S

Reprinted from *Student and Intellectual Observer.*

VIRGIL (Publius Vergilius Maro) 323 1872

The bee. Virgil's Fourth *Georgic* in rhyming verse by Francis David Morice./Oxford, Shrimpton, also London, Whittaker, 1872./30 p./A

Verse.

BUSCH, Wilhelm 324 [1872]
Buzz a buzz, or, The bees, done freely into English from the German . . . by the author of *My bee book* [W. C. Cotton]. / London, Griffith & Farran, also Chester, Phillipson & Golder, [1872]. / vi, 77 p. illus. / A G S . M

Anr. ed. / [1872]. / A G
Anr. ed. / [1890]. / A

Verse, for children. / Hodgson 373.
Pub. in U.S.A. translated by Hezekiah Watkins. / 1873. / Johansson 77.

Busch was a well known comic author and wrote many children's books. The original German title is *Schnurrdiburr, oder die Bienen.*

WOOD, Rev. J[ohn] G[eorge], M.A., F.L.S. 325 1872
Insects at home: being a popular account of British insects, their structure, habits and transformations. / London, Longman, 1872. / illus. / Bees pp. 368–380. / A

Anr. ed. / London, Longmans Green, 1892. / xx, 670 p. illus. XX pl. col. front. / Bees pp. 348–380. / B

A BEEKEEPER 326 1873
To beekeepers—a few plain directions how to take honey without killing the bees, by a beekeeper. / London, Elliott Stock, 1873. / 8 p. / A

CHESHIRE, Frank R. 327 [1873]
Practical bee-keeping: being plain instructions to the amateur for the successful management of the honeybee. / London, 'The Bazaar' Office, [1873]. / 46 p. illus. / A G S . M

2nd ed. rewritten and enlarged by F. Cheshire. / London, 'The Bazaar' Office, [1878]. / 117 p. illus. / A B G S . M W
Anr. ed. / London [1886]. / . M

The 1st ed. is not listed under Cheshire in A's catalogue, presumably on the assumption that he was responsible only for rewriting and enlarging the 2nd ed. Fraser's *History* (p. 88) says that the A copy is undated but bears the stamp 8. De. 73. Thus 1873 must be the correct date, although others give the date 1874. We have used the date 1878 for the 2nd ed. as given in A's catalogue, although Walker p. 13 and Moir p. 59 give 1876.

PALLADIUS, Rutilius Taurus Emilianus 1873
Palladius on husbondrie. (*See ms* 12)

328 1874

Who was the first architect? or, Bees and bee-hives./London, Nelson, 1874./72 p. illus./A

Anr. ed./London, Nelson, 1883./72 p./S

Anr. ed./London, Nelson, 1885./72 p. illus. front./B

For children./Hodgson 372.

LUBBOCK, John (Baron Avebury), M.P., F.R.S., D.C.L., LL.D. 329 1874

On the origin and metamorphoses of insects./London, Macmillan, 1874./xvi, 108 p. illus./A

Anr. ed./London, Macmillan, 1890./xvi, 108 p. 2 leaves, illus. incl. front. 6 pl./A B S

Macmillan's Nature Series.

LUBBOCK, John (Baron Avebury), M.P., F.R.S., D.C.L., LL.D. 330 1875

On British wild flowers considered in relation to insects./London, Macmillan, 1875./186 p. illus./A

Subsequent eds. 1875, 1882, 1903./B S

Macmillan's Nature Series.

COCKBURN, A. 331 1875

The Scotch bee-keeper: a treatise on the proper management of bees./Huntley (Aberdeenshire), Author, 1875./35 p./G S . M

332 1875

Half hours in the tiny world: wonders of insect life./London, Isbister, 1875./illus./Bees and bee hives pp. 49–81./A

Many reprints 1878, 1880, 1896./A S

Half Hour Library.
For children./Hodgson 370.

HUNTER, John 333 1875

A manual of bee-keeping./London, Hardwicke, 1875./vii, 231 p. illus. front./A B G . M

2nd ed./London, Hardwicke, 1875./ix, 1 leaf, 239 p. illus. front./B R . M
Reprinted 1876./London, Hardwicke & Bogue, 1876./B S
3rd ed. rev. and enl./London, Bogue, 1879./viii, 220 p. illus./A B G S . M W

4th ed. / London, Allen, 1884. / viii, 220 p. illus. / S . M
This ed. is a reprint of the 3rd ed.

Hunter was Hon. Sec. of the B.B.K.A.

MICHELET, Jules 334 1875
The insect. / London, Nelson, 1875. / Bees pp. 293–329./illus. by Giocomelli. / A
Anr. ed. / 1883. / A S

Translated from French by W. H. D. Adams. There are almost 150 engravings (including several of bees) which are described in the preface as 'exquisite'. Michelet was known for four works, *L'oiseau* (1856), *L'insecte* (1857), *La mer* (1861) and *La montagne* (1868).

NEIGHBOUR, George, & Sons 334a 1875
Neighbour's improved bee-hives for taking honey without the destruction of the bees. / London, [Kent & Neighbour], 1875. / 31 p. illus.

RUSBRIDGE, Alfred 335 1875
A book for bee-keepers, showing how anyone so disposed may, in a fair locality, with good management, realise net profit averaging from 100 to 300 per cent, annually on the capital invested, in keeping bees. / London, Allen, 1875. / x, 1 leaf, 63 p. illus. / A G S . M

He made and sold hives and equipment, and the book is in effect an illustrated catalogue, with added descriptions. There is also much Sussex bee-lore.

ABBOTT, Charles Nash 336 1876
Beekeeping for the million. / London, British Bee Journal, 1876. / 75 p. / B G S . M

Leaflets for cottagers, condensed from *B.B.J.*
Abbott was the first editor of *B.B.J.* from 1873 to 1894.

DANYELL, Arthur Johnson 337 1876
The Italian system of bee-keeping: being an exposition of Don Giotto Ulivi's economical frame hives and honey extractor. Edited by W. B. Tegetmeier. / London, Field Office, 1876. / 24 p. illus. / A G S . M

DESBOROUGH, J. G. 338 1876
Lecture on bee-keeping. / Hanwell (Oxon.), British Bee Journal, also London, Houlston, also Stamford (Lincs.), Johnson, 1876. / 19 p. / A B G . M

2nd ed./Stamford (Lincs.), for Herts. Beekeepers' Association, 1879./ 21 p./G S . M

GLINN, Mrs. Helen 339 1876

Beeswing's advice to bee-keepers./Hereford, reprint from *Hereford Times*, 1876./24 p. illus./H . M

3rd ed./Hereford, 1877./1 leaf, 28 p. illus./G H

4th ed./Hereford, 1878./1 leaf, 28 p. illus./. M

EDGEWORTH, M[ichael] Pakenham, F.L.S., F.A.S. 340 1877

Pollen./London, Hardwicke & Bogue, 1877./92 p. 24 pl./A S

2nd ed. rev. and corr./London, Hardwicke & Bogue, 1879./3 leaves, 98 p. XXIV pl./A B

MARTIN, Piers Edgcumbe 341 1878

The bee-keeper's almanac, containing prognostications of weather expected and what to do in bee management./London, Simpkin Marshall, also Winchester (Hants.), Tanner, 1878./16 leaves./A . M

Anr. ed./London, Simpkin Marshall, also Winchester (Hants.), Tanner, 1879./16 leaves./A S . M

Anr. ed./London, Simpkin Marshall, also Winchester (Hants.), Tanner, 1880./16 leaves./A G

Anr. ed./London, Simpkin Marshall, also Winchester (Hants.), Tanner, 1881./80 p./A

MARTIN, Piers Edgcumbe 342 1878

The great Hampshire bee farm near Stockbridge by P. E. Martin proprietor, and manager. Inventor of the Sailor's beehive, now the standard bar-frame hive for the United Kingdom. Its principles and method of working./London, Simpkin Marshall, 1878./20 p. illus./A . M

Walker's copy is bound with Martin's *The bee-keeper's almanac* for 1878 and 1879.

'The standard bar-frame hive or Sailor's bee hive specially adapted for doubling and sectional super principle . . . Choice honey in five pound tins sent to any address carriage paid on receipt of P.O. for 6/6.' Martin described himself as a master mariner. *See* Fig. 14.

TITE, C. 343 1878

A lecture on bee-keeping./Southall (Middx.), 1878./20 p.

OPEN TO ALL VISITORS, SUNDAYS EXCEPTED.

THE GREAT HAMPSHIRE BEE FARM,

NEAR STOCKBRIDGE.

THE STANDARD BAR-FRAME HIVE,

OR

'SAILORS' BEE-HIVE.'

Specially adapted for the DOUBLING and SECTIONAL Super principles.

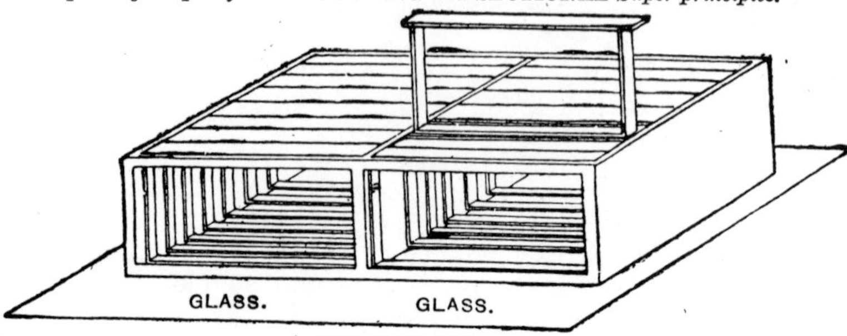

SECTIONAL SUPER.

HIVE SINGLE AS IN WINTER,
AND WITH SECTIONAL SUPER ON.

HIVE DOUBLED.

Sent out complete on either principle, waxed and painted ... 30s.
Packing and Crate ... ... 5s.

2s. allowed on return of Crate.

SWARMS are sent to all parts of the United Kingdom safely every Season in the 'Sailors' Bee Hive.'

CHOICE HONEY ALWAYS ON HAND.

BUCKLEY, Arabella Burton 344 1879
The fairyland of science./London, Stanford, 1879./Bees pp. 193–237./illus./A

Reprinted 1881, 1892./B (1881).
Anr. ed./London, Macmillan, 1903.
Reprinted 1904.
Reprinted with additions and corrections 1906.
Reprinted 1909, 1913, 1919, 1922, 1929./B (1929)

A series of lectures for children./The date 1829 given by Hodgson (410) must be wrong.

GREIG, R. M. 344a 1879
Bees and bee-keeping: a short treatise on apiculture, for the guidance of amateurs./[Aberdeen, Author, 1879]./67 pp.

Known only from the Royal Library, Windsor Castle, which has a copy of the book presented to Queen Victoria, with a handwritten inscription 'Will her Most Excellent Majesty be Graciously pleased to accept from a young author.' The work originally appeared as a series of articles in the *Daily Free Press*.

LINDSAY, William Lauder 345 1879
Mind in the lower animals in health and disease./London, Kegan Paul, 1879./Bees Vol. 1 pp. 134–135, etc./A S

RIBEAUCOURT, C. de 346 1879
A manual of rational bee-keeping./London, Bogue, 1879./xv, 108 p. catalogue, illus. front./

Anr. ed./London, Diprose & Bateman, [n.d.]./Apparently identical except for title-page./A B G S . M W

Study of Bogue's 32-page catalogue suggests that the Diprose & Bateman ed. is the earlier.
Translated by Arthur F. G. Leveson Gower from *Manuel d'apiculture rationelle*.

Fig. 14. Piers Edgecombe Martin, as proprietor and manager, wrote a book on *The great Hampshire bee farm* (342) in 1878. The advertisement shown here appeared in 1881, in E. Bartrum's *The Stewarton: the hive of the busy man* (entry 355), a book whose frontispiece has often been reproduced, e.g. by H. M. Fraser (635).

VIRGIL (Publius Vergilius Maro) 347 1879

Virgil's Georgics Book IV./Dublin, Ponsonby, 1879./21 p./A

Translated by S. Percy Reed.

BAYNES, F. H. 347a 1880

Myrmecologia. Bound notebook containing a handwritten manuscript, with a label on the cover in L. E. Snelgrove's handwriting: 'Myrmecologia. Manuscript of a learned Dissertation on Ants by F. H. Baynes, 1880. Unpublished. Occasional references to Bees.'/B

Presented by Mr. Snelgrove. We know nothing of the author or the history of the manuscript.

BÜCHNER, Ludwig 348 1880

Mind in animals./London, Freethought, 1880./illus./*The bee nation* pp. 212–289./A G S

Anr. ed./London, Freethought, 1903./xii, 359 p./*The bee nation* pp. 211–289./. M

Translated from the German (3rd ed. 1880) by Annie Besant.
International Library of Science and Freethought Vol. 1.

BRITISH BEE-KEEPERS' ASSOCIATION 349 1880

Modern bee-keeping: a handbook for cottagers published for British Bee-Keepers' Association./London, Low, 1880./72 p. illus./A G . M

2nd ed. rev. and enl./London, Longman Green, 1881./viii, 80 p. illus./A G . M

3rd ed. rev. and enl./London, Longman Green, 1882./viii, 9–84 p. illus./A G . M

4th ed. rev. and enl./London, Longman Green, 1883./viii, 9–90 p. illus. incl. front./A B G S

5th ed./London, Longman Green, 1884./viii, 9–90 p. illus. incl. front./A B S

6th ed. rev. and enl./London, Longman Green, 1886./viii, 9–94 p. illus. incl. front./A B G S

7th ed. rev. and enl./1890./93 p. illus./A . M

8th ed. rev. and enl./1895./95 p. illus./A P

9th ed. rev. and enl./London, Longman Green, 1903./viii, 9–103 p./A S . M

9th ed. reprinted./1904./viii, 9–103 p. illus.

The frontispiece is shown in Fig. 15.

HIVING UNDER DIFFICULTIES.

Fig. 15. Frontispiece of *Modern bee-keeping* published by the B.B.K.A. in 1880 (entry 349). Later illustrations in the book show 'modern' frame hives, but the frontispiece is of interest as a social document. The lady of the house is tanging the bees by hitting a spoon against a dustpan (this was supposed to make the swarm settle). It is the servant girl who climbs up the ladder to dislodge the swarm, and a farmhand who holds a skep to catch the falling bees. No one wears a bee veil.

CARR, William Broughton 350 1880

Introduction, or, Early history of bees and honey. / Salford (Lancs.), Roberts, 1880. / 16 p. / G

Carr was a well known writer, and joint editor with Cowan of the *Bee-keepers Record.* In 1890 he described his hive in an article in the journal signed W.B.C. The W.B.C. hive is still in use in England ninety years later.

351 1880

Jenny and the insects; or, Little toilers and their industries. / London, Nelson, 1880. / 267 p. illus. by Giacomelli incl. front. / *The little confectioners* pp. 55–102. / A B S

For children / Hodgson 364.

RAYNOR, Rev. George, M.A. 352 1880

Queen introduction. The Ligurian queen bee; her introduction to alien stocks, and the best means of pure propagation. / London, British Bee-keepers' Association, 1880. / 24 p. illus.

2nd ed. / London, British Bee-keepers' Association, 1884. / 24 p. illus. / B S

A paper read at a quarterly meeting of the British Bee-keepers' Association on 14 January 1880.

ROBINSON, James F. 353 1880

British bee-farming: its profits and pleasures. / London, Chapman & Hall, 1880. / xiii, 1 leaf, 206 p. illus. / A B G S . M W

2nd ed. / London, Chapman & Hall, 1889. / xiii, 1 leaf, 206 p. illus. / B S . M

ADDEY, Thos. 354 [1881]

Bee-keeping for amateurs: being a short treatise on apiculture, on humane and successful principles. / London, 'The Bazaar' Office, [1881]. / 23 p. / A G S . M

Anr. ed. or reprint with advertisements dated 1882. / 23 p. 10 leaves. / B

Addey was called the Lincolnshire Apiarian.

BARTRUM, Rev. Edward 355 1881

The Stewarton: the hive of the busy man. / London, Longman Green, 1881. / 45 p. illus. / A B G S . M

A paper read at a meeting of the British Bee-keepers' Association in October 1880, with an account of the discussion which followed. Bartrum was Headmaster of King Edward VI School, Berkhamsted, and a member of the Committee of the B.B.K.A. Fig. 14 shows an advertisement in this book.

COWAN, Thomas William, F.G.S., F.R.M.S. 356 [1881]
British bee-keeper's guide book to the management of bees in moveable comb hives, and the use of the extractor. / London, Houlston, [1881]. / viii, 135 p. illus. / A G P S . M

2nd ed. / London, Houlston, 1882. / viii, 135 p. / B G . M W
We believe that this is a reprint of the 1st ed.
3rd ed. rev. enl. / London, Houlston, [1883]. / xii, 160 p. illus. / B S . M
4th ed. / London, Houlston, [1884]. / xii, 164 p. illus. / . M
5th ed. rev. / London, Houlston, [1884]. / xii, 164 p. illus. / B
6th ed. / London, Houlston, [1885]. / xii, 164 p. illus. / B G
7th ed. rev. / London, Houlston, [1886]. / xii, 164 p. illus. / B G S
8th ed. rev. / London, Houlston, [1886]. / xii, 164 p. illus. / B G S
9th ed. rev. enl. / London, Houlston, [1888]. / xii, 174 p. illus. / B G . M
10th ed. rev. enl. / London, Houlston, [1889]. / xii, 174 p. illus. / B G S
11th ed. rev. enl. / London, Houlston, [1891]. / xii, 186 p. illus. / B G
12th ed. rev. / London, Houlston, [1893]. / xii, 188 p. illus. / B
13th ed. rev. / London, Houlston, [1894]. / xii, 188 p. illus. / B
14th ed. rev. enl. / London, Houlston, [1896]. / xii, 176 p. illus. / B G S . M W
15th ed. rev. / London, Houlston, [1898]. / x, 176 p. illus. 1 pl. / B G . M
16th ed. rev. enl. / London, Houlston, [1900]. / x, 180 p. illus. 1 pl. / B G . M
17th ed. / London, Houlston, [1901]. / x, 180 p. illus. 1 pl. / B . M
18th ed. / London, Houlston, [1903]. / x, 180 p. illus. 1 pl. / B G . M W
19th ed. rev. enl. . . . moveable-comb hives, and the use of modern bee-appliances. / London, Madgwick, Houlston & British Bee Journal, [1907]. / viii, 226 p. illus. 1 pl. front. / B G . M
20th ed. / London, Madgwick, Houlston & British Bee Journal, 1911. / viii, 226 p. illus. 1 pl. front. / B G S
Described as Coronation Edition.
21st ed. / London, Madgwick, Houlston & British Bee Journal, 1913. / viii, 226 p. illus. 1 pl. front. / B P S
22nd ed. / London, Madgwick & British Bee Journal, 1915. / viii, 226 p. illus. 1 pl. front. / B S
23rd ed. / London, Madgwick & British Bee Journal, 1919. / viii, 226 p. illus. 1 pl. front. / B S
24th ed. / London, Larby & British Bee Journal, 1921. / viii, 226 p. illus. 1 pl. front. / B G . M W
25th ed. / London, Larby & British Bee Journal, 1924. / viii, 226 p. illus. 1 pl. front. / B G
Manley, *Bee-keeping in Britain* p. 423: 'It is largely through the influence of this little guide that the form of British bee-keeping practice took on its well-known characteristics.' The book had almost as much influence abroad, being translated into at least six languages. All eds. have many illustrations.

DAY, John 1881
Works . . . (*see* 29)

MEETKERKE, Cecelia Elizabeth 357 1881
The guests at home./London & Edinburgh, Griffin & Farran, also New York, Dutton, 1881./61 p./A G S

Sequel to *Guests of flowers* (1881). In the first of these two children's books bees are among the guests of flowers, and the sequel describes these guests at home.

SWINTON, Archibald Henry 358 [1881]
Insect variety; its propagation and distribution./London, Cassell, [1881]./326 p./Bees p. 212 etc. illus./A S

VIRGIL (Publius Vergilius Maro) 359 1881
The Georgics of Virgil./[London, Kegan Paul] 1881./122 p./A D

Translated into English verse by James Rhoades.

360 [1882]
Bees, rabbits and pigeons: how to breed and how to rear them on sound principles./London, Ward Lock, [1882]./illus./Bees pp. 1–69./A B G S . M

Country Life Series.

BLOW, Thomas B. 361 1882
A bee-keeper's experience in the East, with discussions upon the merits of Cyprian and Syrian races of bees./London, British Bee-keepers' Association, 1882./22 p./G

Anr. ed. . . . among the queen raisers in the north of Italy and Carniola [and A visit to Mr. Blow's bee garden]/Welwyn, (Herts.), Author, 1887./48 p. illus. front/(B) G . M
Read to the British Bee-keepers' Association 12 April 1882. The later ed. omits the discussion which took place at the meeting.

BRITISH BEE-KEEPERS' ASSOCIATION 362 [1882]
Catalogue of books./London./[1882]./16 p./G

Superseded by 455.

DUNCAN, Peter Martin 363 1882
The transformations (*or metamorphoses*) *of insects*: being an adaption for English readers of M. E. Blanchard's *Metamorphoses, moeurs et instincts des insectes* . . . and a compilation from the works of Newport, Charles Darwin, Spence, Bate, Fitz Muller, Packard, Lubbock, Stanford and

others. / London, Cassell, 1882. / Metamorphoses of Hymenoptera pp. 162–287. illus./A S

DYER, Gertrude P. 364 1882

Elsie's adventures in insect-land. / London, Ward, 1882. / 287 p. / Bees pp. 92–143. illus. / A S

For children. / Hodgson 361.

DZIERZON, Johann 365 1882

Dzierzon's rational bee-keeping; or, The theory and practice of Dr. Dzierzon, of Carlsmarkt. Edited and rev. by Charles Nash Abbott. / London, Houlston, also Southall (Middlesex), Abbott, 1882. / xvi, 350 p. illus. / A B S . M

Later copies with 'second thousand' on title page dated 1882 appear to be identical / A B G

Translated by H. Dieck and S. Stutterd from *Rationelle Bienenzucht*, published in Brieg in 1861 and 1878. Dzierzon discovered parthenogenesis in bees.

LUBBOCK, John (Baron Avebury), M.P., F.R.S., D.C.L., LL.D. 366 1882

Ants, bees and wasps a record of observations on the habits of the social Hymenoptera. / London, Kegan Paul & Trench, 1882. / xix, 448 p. 5 col. pl. / B

Reprinted in various so-called eds. to 17th, 1882–1915. / A B (14th, 16th, 17th). / G S

International Scientific Series Vol. 40.

New ed. edited and annotated by J. G. Myers with 4 new col. pl. by A. J. E. Terzi. / London, Kegan Paul, 1929. / xviii, 1 leaf, 377 p. VI pl.

Published in U.S.A. 1882. Johansson 41.

Lubbock had intended to make observations principally on bees, but found ants more convenient; only pp. 274–310 (and none of the pls.) deal with bees.

ROMANES, Georges John 367 1882

Animal intelligence. / London, Kegan Paul, 1882. / illus. / A

3rd ed. 1883. / Bees and wasps pp. 143–197. illus. / A

8th ed. 1904. / G / 9th ed. 1910. / A

International Scientific Series Vol. 41.

MORWOOD, Vernon S. 368 1882
Facts and phases of animal life, and the claims of animals to humane treatment./London, J. Hogg, 1882./Buzzings from a beehive pp. 32–39./A S

SIMMINS, Samuel 369 [1882]
Facts for bee-keepers. The Simmins method of direct introduction adapted to every condition under which it is necessary to give an alien queen to a colony of bees . . . also other practical hints about queens, and uniting in general./London, [Author], [1882]./28 p./A B G S . M

ALDRIDGE, J. M. 370 1883
On bees and beekeeping./Dublin, Hodges, Figgis, 1883./Bees pp. 23–68./A
In the Report of the Joint Dairy Show Committee (1882).

L., E. R. 371 [1883]
The cottage bee-keeper./Norwich, Norwich Temperance Tracts, [1883]./8 p./G

LUBBOCK, John (Baron Avebury), M.P., F.R.S., D.C.L., LL.D. 372 [1883]
Chapters in popular natural history; . . . for use in schools./London, National Society, [1883]./Bees pp. 63–73. illus./A S

MÜLLER, Hermann 372a 1883
The fertilisation of flowers./London, Macmillan, 1883./x, 1 leaf, 669 p. illus.
Translated by D'Arcy W. Thompson from *Die Befruchtung der Blumen durch Insekten* . . . (1873). Major work with 3748 references to publications on pollination, and much information on bees.
See also Knuth (435).

ROMANES, Georges John 373 1883
Mental evolutions in animals: with a posthumous essay on instinct by Charles Darwin./London, Kegan Paul, 1883./Bees pp. 166–168 etc./A S
The book is a continuation of *Animal intelligence* (367). It incorporates some unpublished material by Darwin, including the full text of the part of his chapter on instinct that was written for *Origin of species* but omitted from the published version. There are numerous references to bees.

RUSBRIDGE, Alfred 374 1883
Bee-keeping, plain and practical; and how to make it pay./London, Allen, 1883./143 p. illus./A B S G S . M W

PAN *pseud.* **(THOMSON, William)** 375 1883
An essay of bees; being the experience and observations in his own apiary, with hints on their natural history./Glasgow, by Nimo for Pan, 1883./ 84 p. illus./S . M

Anr. ed./Glasgow, Pan, [1884]./84 p. illus./G

Prize essay, Highland & Agricultural Society of Scotland, 1882. Thomson calls himself A Lanarkshire Bee-keeper.

BURROUGHS, John 376 1884
Locusts and wild honey./Edinburgh, D. Douglas, 1884./The pastoral bees pp. 11–43./A G

Pub. in U.S.A. 1879./Johansson 74.

CHESHIRE, Frank R., F.R.M.S. 377 1884
Foul brood (not micrococcus, but bacillus), the means of its propagation and the method of its cure./Kings Langley (Herts.), British Bee-Keepers' Association, 1884./48 p. illus./B G

COWAN, Thomas William, F.G.S., F.R.M.S., etc. 378 [1884]
British bee-keepers' practical note book: followed by rules for the general management of moveable comb hives throughout the year./Kings Langley (Herts.), Huckle, [1884]./56 p./G S

2nd ed./London, Houlston & British Bee Journal, 1904./73 p./A G
3rd ed./London, Madgwick, Houlston & British Bee Journal & Bee-keepers' Record Office, [1908]./72 p. illus./B G P S

[DAWSON, A. G.] 379 1884
The amateur's guide in bee-keeping./Macclesfield (Ches.), A. G. Dawson, 1884./50 p. illus./A G S

HARRIS, W[illiam] H[etherington], B.Sc. 380 [1884]
The honey-bee: its nature, homes and products./London, Religious Tract Society, [1884]./xv, 272 p. illus. incl. front./A B G S . M W

HEHNER, Otto, F.C.S., F.I.C. 381 1884
Adulteration of honey./Kings Langley (Herts.), British Bee-Keepers' Association, 1884./36 p./B . M

Paper read during the International Health Exhibition, at a Conference of the B.B.K.A., 1884.

SIMMINS, Samuel 381a 1884
Foul brood. Paper read at a quarterly meeting of the British Bee-Keepers' Association, on 16 January 1884./Kings Langley (Herts.), B.B.K.A., 1884./23 p./G

NOEL, Maurice 382 [1885]
Buz: or the life and adventures of a honey bee./Bristol, Arrowsmith, [1885]./5 leaves, 140 p. illus. 21 cm./A B G . M

2nd ed./Bristol, Arrowsmith, also London, Simpkin Marshall, [1885]./140 p./G S

For children./Hodgson 358.
Pub. in U.S.A. 1886./Johansson 387.
Walker p. 52: 'Front. by Linley Sambourne, reproduced in colour on outside b[oar]ds.'

(*cancelled*) 383

WOOD, Rev. J[ohn] G[eorge] 384 1885
Half hours with a naturalist./London, Isbister, 1885./Hive bees pp. 204–234. illus./A

SCHOLES, S. E. 385 [c. 1885]
Three little folks who mind their own business; or The bee, the ant, and the spider./London, Woolmer, [c. 1885]. 93 p. 3 pl. illus. 12 cm./The bee pp. 9–39 illus.
For children.

[ANDERSON, Charles G.] 386 [1886]
The bee keepers alphabet./London, Simpkin Marshall, [1886]./15 p. illus. 18 × 25 cm./A B . W

Hodgson 355.
Rare and quaint paper-covered book, each letter being illustrated. The author was Hon. Sec. of Somerset Beekeepers' Association

BLOW, Thomas B. 387 1886
Bees, hives and honey: illustrated catalogue of beekeepers' supplies./Welwyn (Herts.), Thomas B. Blow, 1886./60 p. illus./A(B) G

CHESHIRE, Frank R., F.L.S., F.R.M.S. 388 1886
Bees and bee-keeping; scientific and practical: a complete treatise on the anatomy, physiology, floral relations and profitable management of the hive bee./London, Upcott Gill:

Vol. I Scientific, 1886./viii, 1 leaf, 336 p. illus. VIII pl. incl. front./ A B G S . M W
Vol. II Practical, 1888./3 leaves, 652 p. illus. front./A B G S . M W
Anr. ed./London, Bazaar, Exchange & Mart, 1921./2 vols. illus./A B G S
Reprint with appendix of 84 p. by J. B. Lamb.

Some copies of 1st ed. have no date on title page and are presumably reissues. B has the author's interleaved copy of Vol. I of the 1st ed. The book was issued in parts, which may account for some of the variations. Manley, *Bee-keeping in Britain* p. 423: '. . . in its day the leading work on anatomy published in Britain'.

DOBBIE, Henry 389 [1886]
Bee pasturage: giving observations and practical instructions on the cultivation and propagation of the best honey and pollen producing plants./London, Jarrold, [1886]./75 p./G S . M
Preface dated 6 April 1886.

JENYNS, Rev. [Charles] F[itzgerald] G[ambier] 390 1886
A book about bees: their history, habits, and instincts; together with the first principles of modern bee-keeping for young readers./London, Wells Gardner Darton, 1886./xxiv, 200 p. illus. incl. front./A B G S . M W
Anr. ed./London, Gardner, 1888./xxiv, 200 p. illus. incl. front./B S . M
For children./Hodgson 357.
Fig. 16 shows two of the engravings.

SIMMINS, Samuel 391 1886
Simmins' original non-swarming system: as adapted to hives in present use./Brighton (Sussex), Author, 1886./viii, 9–62 p., 2 leaves. illus./B G
Anr. ed./1888./64 p. illus./. M
Title page: 'A new era in modern bee-keeping.'

CHESHIRE, Frank R., F.L.S., F.R.M.S. 392 1887
The apparatus for differentiating the sexes in bees and wasps: an anatomical investigation into the structure of the receptaculum seminis and adjacent parts./London, Clowes, 1887./15 p. 2 pl./A G S . M
Walker p. 24: '*Ex*. Journ. R. Microscop. Soc. Feb. 1885'.
Moir Cat. p. 29 says: 'Repr. from Jnl. of Roy. Microscop. Soc., Dec., 1884', and gives date 1884 to book.

COWAN, Thomas William 393 1887
Doubling and storifying for extracted and comb honey and the prevention of swarming./Kings Langley (Herts.), Huckle, 1887./15 p./G

A Bee Tent.

The Cottager and his Bees.

HAZLITT, William Carey 394 1887

Gleanings in old garden literature./London, Elliott Stock, 1887./Bees pp. 48–53, 130, 203–213/A S

Book-lover's Library.

SIMMINS, S[amuel] 395 1887

A modern bee-farm and its economical management. Showing how bees may be cultivated as a means of livelihood; as a health-giving pursuit; and as a source of recreation to the busy man. Profits made certain by growing crops yielding the most honey, having also other uses; and by judgment in breeding a good working strain of bees./Rottingdean (Sussex), S. Simmins, also London, Pettitt, 1887./1 leaf, xiv, 195 p. illus. 4 pl. incl. front./B G S . M W

There was also a cheaper ed. in limp cloth cover without pls.

Rev. ed./1891./G (part 3 pp. 33–48)

New and rev. ed./London, Woodford Fawcett, 1893./xiv, 274 p. illus./ A B G S . M

Rev. ed. *£300 per annum from 30 acres*, or a modern bee-farm . . . also notes upon profitable grass-farming and dairying and orchard planting/ London, Woodford Fawcett, 1904./xii, 405 p. illus. 11 pl./A B G S . M

Anr. ed. rev. *A modern bee-farm* . . , grass-farming and dairying . . . etc./ Heathfield (Sussex), Author, 1914./xvi, 479 p. illus. 12 pl./B S . M W

Rev. ed. *A modern bee-farm and its economic management.*/Heathfield (Sussex), Author, 1928./xii, 506 p. illus. 11 pl./A B G

The plates in the 1st ed. were original sepia photographs of the author's 'domestic hive' (his family), his home apiary, the covered apiary, and queen and workers life size on a comb. These were omitted from the cheaper ed. but were available separately. A revised ed. was being produced in parts, of which G has Part 3 only. The text differs from pp. 33–48 in the 1st ed. and the new and revised ed., 1893, but we have not seen any other copies complete or otherwise. If this ed. was completed in 1892 it would account for the fact that the new and revised ed. contains an

---

Fig. 16. Two of the engravings from the Rev. F. G. Jenyns' *A book about bees* (1886, entry 390). The upper one (from page 199) shows the 'bee-tent' where, 'safe from attack behind the gauze net, many will be gathered to witness some expert engaged in driving bees'. The charming picture below (from page 46) shows the cottager with his bees; he has presumably transferred a colony to a movable-frame hive and is able to see for the first time what goes on inside. There seem to be no flying bees to trouble him or his watching family.

'Introduction to Revised (1892) Edition', although all copies of the new and revised ed. that we have seen are dated 1893 on the title page.

The 1904 ed. deals also with crops, poultry and livestock to be managed in conjunction with the bee farm, with plates of hens and cows. The 1914 ed. mentions on the title page, 'New and important information relating to the treatment and cure of bee paralysis, or the Isle of Wight disease, including the author's definite law for ensuring the decline of that malady'. This ed. contains a plate of a typical Jersey cow but the emphasis on other animals is very much reduced.

The 1914 and 1928 eds. state that the date of the 1st ed. was 1888, but 1887 is correct. B's copy of the 1928 ed. is the one used by the author in making notes for a proposed new edition, never published.

BATH, W[illiam] Harcourt 396 1888

The young collector's handbook of ants, bees . . . and flies. / London, Swann Sonnenschein Lowry, 1888. / illus. front. / *Order Hymenoptera* pp. 13–33. / A B G S

2nd ed. / London, 1890. / illus. / S
5th ed. / 1912.

Young Collector Series.
For children. / Hodgson 354.

DAY, John 1888

The parliament of bees. (*see* 29)

397 1888

Honey for money. / Dublin, Freeman's Journal, 1888. / 48 p. / A
Freeman's Handbook No. 1.

HOOKER, John M. 398 [1888]

Hooker's guide to successful bee-keeping. / Kings Langley (Herts.), John Huckle, [1888]. / viii, 62 p. illus. / S . M

JONES, H. P. & JONES, Michael D. 399 1888

Y gwenynydd [The beekeeper]: sef llaw-lyfr ymarferol ar gadw gwenyn. / Bala (Merioneth), H. Evans, 1888. / viii, 9–127 p. illus. front. / B

The first bee book in Welsh, according to the author in his dedication.

LUBBOCK, John (Baron Avebury), M.P., F.R.S., D.C.L., LL.D. 400 1888

On the senses, instincts and intelligence of animals with special reference to insects./London, Kegan Paul, 1888./Bees pp. 194–201. illus./A S

International Scientific Series Vol. 65.

WEBSTER, W. B. 401 1888

The book of bee-keeping: a practical and complete manual on the proper management of bees. Specially written for amateurs, beginners, cottagers and others who are desirous of keeping two or three hives of bees so as to yield the best results./London, Upcott Gill, 1888./2 leaves, 103 p. illus./ A G S . M W

2nd ed. . . . specially written for . . . others who are not experts./London, Upcott Gill, also New York, Scribner, 1901./3 leaves, 103 p. illus./A
Other copies of 2nd ed. 3 leaves, 103 p. illus./?earlier; London, Upcott Gill (170 Strand)./B
?later: London, Upcott Gill (Bazaar Buildings, Drury Lane), also New York, Scribner./S
Anr. 2nd ed./London, Upcott Gill, also New York, Scribner, 1905./ 2 leaves, 104 p. illus.
4th ed./London, Upcott Gill, also New York, Scribner, 1908./2 leaves, 104 p. illus.
Anr. 4th ed./London, Upcott Gill, [1910]./3 leaves, 104 p. illus./A B S
5th ed. not traced.
6th ed./London, The Bazaar, Exchange & Mart, [1929]./3 leaves, 104 p. illus./B S
7th ed./London, The Bazaar, Exchange & Mart, [1938]./vii, 96 p. illus./ B S
8th ed. rev. by J[oseph] Herrod-Hempsall./London, The Bazaar, Exchange & Mart, [1947]./89 p. illus./A S
Pub. in U.S.A. 1901./Johansson 588.

DENNLER, J. 402 1889

Bees-wax; its economical uses and conversion into money./Kings Langley (Herts.), 1889./19 p. illus./G S . M

Translated from German by T. W. Cowan.

BANCKS, Rev. [Charles] Gerard W[instanley], M.A. 403 [c. 1890]

Honey and its uses./Dartford (Kent), Perry, [c. 1890]./12 p./G . M

Rev. ed./Dartford (Kent) & London, Perry, [1898]./12 p. 14 cm./A B

COWAN, T[homas] W[illiam], F.L.S., F.G.S., F.R.M.S., F.S.Sc. 404 [1890]
The honey-bee: its natural history, anatomy, and physiology. / London, Houlston, [1890]. / xi, 220 p. illus. / A B S . M W

2nd ed. / London, Houlston, [1904]. / xi, 220 p. illus. incl. front. / A B G . W

B has translations in German (1891) and Russian (1903).

Ceo est hosebonderie. (*See ms* 7) 1890

LETT, Rev. H. W. 405 1890
Beekeeping for Ulster. / Belfast, 1890. / 20 p.

LOVELL, Kate R. 405a 1890
Nature's wonder-workers being some short life-histories in the insect world. / London, Cassell, 1890. / 285 p. illus. / A

Anr. ed. 1896. / viii, 285 p. illus. / Bees pp. 61–91, illus. / For children.

S., J. *pseud.* **(WALLRAND-EVANS, J. H.)** 406 1890
Bee-keeping; its excellence and advantages. / London, British Bee Journal Office, 1890. / 19 p. / G S

ATKINSON, J[ohn] C[hristopher], D.C.L. 407 1891
Forty years in a moorland parish: reminiscences and researches in Danby in Cleveland. / London, Macmillan, 1891. / Bee customs and notions pp. 126–132. / A(B)

Anr. ed. / London, Macmillan, 1907. / A

SAMSON, G[eorge] Gordon 408 1892
Bees for pleasure and profit: a guide to the manipulation of bees, the production of honey, and the general management of the apiary. / London, Crosby Lockwood, 1892. / 1 leaf, viii, 82 p. illus. / A G S . M

2nd ed. / London, Crosby Lockwood, 1901. / 82 p. illus. / . W
Weale's Scientific and Technical Series.

3rd ed. rev. enl. with a new chapter on beekeeping in hot climates. / London, Crosby Lockwood, 1907. / xi, 100 p. illus. / B G S . M

5th ed. rev. enl. / London, Crosby Lockwood, 1921. / xii, 122 p. illus. / A B S . W
Lockwood's Garden Manuals.

SIMMINS, S[amuel] 409 [1893]
Rapid increase in early spring; or The true principles of stimulation and brood production: together with a revised treatise on successful wintering./London, Woodford Fawcett, [1893]./16 p./B G

SIMMINS, S[amuel] 410 1894
Simmins' new queen-rearing; together with his direct introduction of queens, uniting bees, nucleus swarming, and description of the various races of bees suitable for general use./London, Woodford Fawcett, 1894./32 p./B

WELLS, G. 411 [1894]
Guide book pamphlet on the two queen system of bee-keeping./Snodland (Kent), Gay, [1894]./15 p./A

Anr. ed./Aylesford (Kent), [1896]./. M

Wells's system made a great stir and was in use for several years.

COOK, A[rthur] B[ernard] 412 1895
The bee in Greek mythology./*Journal of Hellenic Studies* 15: 2–25 (1895)./26 cm./B

GILLIES, J. M. 413 1895
Stepping-stones to bee-keeping./Dublin, Gill, 1895./36 p. front./bibliog./B G S . M

WHITE, Charles Nettleship 414 [1895]
Pleasurable bee-keeping./London, Arnold, [1895]./184 p. illus./A B G S . M

BANCKS, Rev. [Charles] Gerard W[instanley], M.A. 415 1895
Mead, and how to make it./Dartford (Kent), Perry, 1895./15 p. 15 cm./A B G . M

Anr. ed./[1898]./15 p./A
4th ed./[1906]./15 p./A
5th ed./Dartford (Kent) & London, Perry, [1907]./15 p. 14 cm./B

BANCKS, Rev. [Charles] Gerard W[instanley], M.A. 416 1896
The production of vinegar from honey./Dartford (Kent), Perry, 1896./19 p./A O

3rd ed./London, 1898./15 p./A . M
4th ed./Dartford (Kent) & London, Perry 1905./15 p. 15 cm./A B

SAUNDERS, Edward, F.L.S. 417 1896
The Hymenoptera Aculeata of the British Islands: a descriptive account of the families, genera, and species indigenous to Great Britain and Ireland, with notes as to habits, localities, habitats, etc. / London, Reeve, 1896. / xii, 391 p. 3 pl. /(+3 leaves interspersed) / A B S

BARTRUM, Rev. Edward
& McCLELLAND, Rev. R. 418 1897
Bees in a bar-frame hive. / London, Society for Promoting Christian Knowledge, 1897. / 16 p. / A S . M W

Helpful Hints for Hard Times.

CRAIG, Annie 418a [1899]
Nora's queer lesson books. / London, Culley, [1899]. / 128 p. 1 pl. front. / Bees pp. 38–61. / A

For children. Nora and Queenie stay with their aunt in the country and receive instruction from lesson books that jumped (a frog), flew (a fly), and stung (a bee), etc. Aunt tells the children how she spent 7/6d on a queen, and shows them the cage in which it arrived by post.

SHARP, David, M.A., M.B., F.R.S. 419 1899
Insects. / London, Macmillan, 1899. / Anthophila or Apidae Bees Vol. 6 pp. 10–70 illus. / A

Reprinted 1901, 1909, 1918.

1918 reprint. xii, 626 p. Bees pp. 10–70 illus. / B

Cambridge Natural History Vol. 6.

1900 to 1926

Beekeeping for the many

O'BRYEN, Turlough B. 420 1900
Instruction in bee-keeping./Dublin, Congested Districts Board for Ireland, 1900./96 p. illus. incl. front./B G
Advocates the C.D.B. hive.

FABRE, J. Henri 421 1901
Insect life: souvenirs of a naturalist./London, Macmillan, 1901./xii, 320 p. 16 pl./Mason bees pp. 271–316.
Reprinted 1913.
Vol. 1 of *Souvenirs entomologiques* translated from French by the author of *Mademoiselle Mori*, preface by David Sharp, M.A., F.R.S., and ed. by F. Merrifield, illus. by M. Prendergast Parker. For the bee chapters in a new translation, *see* 468.

MAETERLINCK, Maurice 422 [1901]
The life of the bee./London, Allen, [1901]./3 leaves, 348 p. bibliog./A B G . M
Reprinted (slight alterations in text) in several so-called editions 1901–1940./356 p./A B G S
Anr. (pocket) ed./London, Allen, 1908./4 leaves, 356 p. 1 leaf, bibliog./A B S
Reprinted 1909–48.
Anr. ed./London, Allen, 1911./vii, 235 p. 13 col. pl. by Edward J. Detmold. bibliog. 28 cm./A B S
Reprinted 1912.
Anr. ed./London, Nash & Grayson, 1929./128 p. bibliog./A B
Nash's Famous Modern Books No. 1.
Pub. in U.S.A. 1901. Johansson 321.
Translated by Alfred Sutro from *La vie des abeilles* (1901).

TAYLOR, E. H. 423 1901
Bee appliances and how to use them./Hull (Yorks.), Burtt, 1901./illus./P S
2nd ed./Hull (Yorks.), Burtt, 1905./146 p. 1 leaf./B
3rd ed./St. Albans (Herts.), Gibbs & Bamforth, 1906./160 p. illus./B S
4th ed./St. Albans (Herts.), 1911./S

5th ed./St. Albans (Herts.), Gibbs & Bamforth, 1918./170 p. 1 leaf, illus./B

CHITTY, Walter 424 1903
Bee-keeping for beginners according to the syllabus of the Board of Education for Schools. / London, Kegan Paul Trench & Trübner, 1903./ viii, 87p. illus./A B G S . M W
For children./Hodgson 347.

HARRISON, Charles 425 1903
The book of the honeybee./London, John Lane, 1903./xii, 132 p. illus. 11 pl. front./A B G S . M W
Handbooks of Practical Gardening XIV.
Hodgson 348./Johansson 210.

REID, John 1903 and 1907
The Scot's gardiner (*see* 59)

SHORTHOUSE, Herbert S., F.C.S. 425a 1903
Profitable bee culture./Birmingham, Cornish, 1903./20 p./G
Text of lecture to the Midland Pharmaceutical Association at the University of Birmingham, 27 Nov. 1902.

STEVENS, Frank 426 1903
Adventures in hiveland./London, Hutchinson, 1903./4 leaves, 227 p. illus. by L. A. Sargent./A B S
For children./Hodgson 350.

427 [1903]
Structure of the bee shown in model./London, Gresham, [1903]./2 p. illus. 18 × 25 cm./A B G
Flat cardboard cut-outs of queen and drone, with lifting flaps; the parts of the bee are numbered and identified. (*See also* 449.)

DAY, John 1904
The parliament of bees (*see* 29)

DIGGES, Rev. J[oseph] G., M.A. 428 [1904]
The Irish bee guide: a manual of modern bee-keeping./Lough Rynn (Co. Leitrim), Irish Bee Journal, also Dublin & Belfast, Eason, also London, Simpkin Marshall Hamilton & Kent, [1904]./viii, 220 p. illus. 20 pl. front./A B G P S . M

2nd ed. *The practical bee guide*: a manual of modern beekeeping. / London, Simpkin Marshall Hamilton & Kent, also Dublin & Belfast, Eason, also Lough Rynn (Co. Leitrim), Irish Bee Journal, [1910]. / ix, 229 p. illus. 2 pl. front. / A B G P S . M W
3rd ed. / London, Simpkin Marshall Hamilton & Kent, also Dublin & Belfast, Eason, also Lough Rynn (Co. Leitrim), Irish Bee Journal, 1917. / ix, 229 p. illus. 2 pl. front. / B
4th ed. / London, Simpkin Marshall Hamilton & Kent, also Dublin & Belfast, Eason, also Lough Rynn (Co. Leitrim), Irish Bee Journal, 1918. / 1 leaf, ix, 229 p. illus. 2 pl. front. / B S
5th ed. / London, Simpkin Marshall Hamilton & Kent, also Dublin & Belfast, Eason, also Lough Rynn (Co. Leitrim), Irish Bee Journal, 1921. / 2 leaves, ix, 229 p. illus. 1 pl. front. / B
6th ed. / London, Simpkin Marshall Hamilton & Kent, also Dublin & Belfast, Eason, also Mohill (Co. Leitrim), Bee Publications, 1928. / 3 leaves, vi, 306 p. illus. front. / B
7th ed. / London, Simpkin Marshall, also Dublin & Belfast, Eason, also Mohill (Co. Leitrim), 1932. / vi, 306 p. vii–xviii, illus. front. / B S
8th ed. / Dublin & Cork, Talbot Press, also London, Simpkin Marshall 1936. / viii, 305 p. illus. front. / B S
9th ed. / Dublin & Cork, Talbot Press, also London, Simpkin Marshall 1941. / viii, 305 p. illus. incl. front. / B S
10th ed. / Dublin & Cork, Talbot Press, also London, Simpkin Marshall 1943. / viii, 305 p. illus. incl. front. / B
11th ed. / 1943.
12th ed. / Dublin & Cork, Talbot Press, 1944. / viii, 305 p.
13th ed. / Dublin & Cork, Talbot Press, 1945. / viii, 305 p. illus. incl. front. / B
14th ed. / Dublin, Talbot Press, 1947. / viii, 305 p. illus. incl. front. / B
15th ed. / Dublin, Talbot Press, 1949. / vii, 249 p. illus. incl. front. / B
16th ed. / 1950 / 256 p.

B has the following eds. 1st, 2nd, 3rd, 4th, 5th, 6th, 8th, 9th, 10th, 13th, 14th, 15th.
The 1st ed. had 20 pls. of which 18 portray prominent Irish beekeepers, but these were not included in subsequent eds. The 8th and subsequent eds. were revised by R. O. B. Manley. Digges was also author of Homestead Readers *Beekeeping*, Dublin, c. 1904, 9 p. illus. / Hodgson 346.

COMSTOCK, Anna Botsford 429 1905
How to keep bees: a handbook for the use of beginners. / London, Doubleday Page, 1905. / 228 p. illus. / A

Pub. in U.S.A. / Johansson 101, Hodgson 344. Many eds.
B has U.S.A. eds. 1914, 1920.

EDWARDES, [E.] Tickner 430 [1905]

Bees as rent payers: a guide for cottagers to the management of a bee-garden. Arundel & Horsham (Sussex), Mitchell, [1905]. / 42 p. / A

New ed. rev. enl. *Bees as rent payers*. A plain practical guide for cottagers to the successful management of a bee garden, and the production and marketing of honey. / Arundel (Sussex), Mitchell, [1906]. / 43 p. illus. 5 pl. / B G S . M

Reprinted from *West Sussex Gazette*.

HASLUCK, Paul N. *editor* 431 1905

Beehives and bee-keepers appliances. / London, Cassell, 1905. / 160 p. illus. / A B G S . M

Many reprints to 1920. B has 1912, 1919, 1920.

Work Handbook Series.

IRELAND. Department of Agriculture & Technical Instruction for Ireland 432 1905

Instruction in bee-keeping for the use of Irish bee-keepers. / Dublin, 1905. / 90 p. illus. / A S . M

Superseded by 677.

New & rev. ed. / London, H.M. Stationery Office, 1912. / 88 p. 4 leaves, 33 pl. front. / A B G S . W

New & rev. ed. / Dublin, Stationery Office, 1931. / 88 p. 40 pl. 1 pull-out pl. / B G

Later eds. 1941, 1943, 1946 (all B).

THE ROYAL FORESTS OF ENGLAND 1905

The Royal forests of England. (*See ms* 9)

SLADEN, F. W. L. 433 1905

Queen-rearing in England and notes on a scent-producing organ in the worker-bee, the honey-bees of India, and enemies of the bee in South Africa. / London, Houlston, 1905. / vi, 1 leaf, 56 p. illus. 1 pull-out pl. front. / A B G S . M

2nd ed. . . . with notes on . . . worker-bee and how pollen is collected by the honey-bee and bumble-bee. / London, Madgwick Houlston, 1913. / viii, 86 p. illus. 2 pl. col. front. / B G S . M W

DUNN, G. C. 434 [1906]

Don'ts and whys in bee-keeping. / London, Drane, [1906]. / 58 p. / A G S . M

KNUTH, Paul 435 1906

Handbook of flower pollination: based upon Hermann Müller's book (*see* 372a). 3 Vol. Oxford, Clarendon Press, 1906./Vol. I. Introduction and literature./xix, 382 p. 1 leaf, illus. front. bibliog. Vol. II. viii, 703 p. Vol. III. iv, 644 p./A B (Vol. I only).

Translated by J. R. Ainsworth Davis. Many references to bees.

[MORLEY, Margaret Warner] 436 1907

The bee people./London, Methuen, 1907./xii, 124 p. illus./A B G S

No author is mentioned on the title page, but a note says that the book is founded on Margaret Warner Morley's book with the same title pub. in U.S.A. 1899.

For children./Hodgson 340.
Many U.S.A. eds./Johansson 354.

EWALD, Carl 437 1907

The queen bee and other nature stories./London, Edinburgh & Dublin, Nelson, 1907./Bees pp. 13–31, illus. 1 pl./A B S

Anr. ed./London, Nelson, 1908./Queen bee pp. 13–31. illus./. W
Anr. ed./London, Nelson, 1930./A

Translated by G. C. Moore Smith from the Danish *Eventyr i udvalg* (1882).
For children./Hodgson 362.
Another of Ewald's books has 11 pages on bees: *The twelve sisters*, translated by Alexander Texeira de Mattos./London, 1923./143 p./B

SAUNDERS, Edward, F.R.S., F.L.S. 438 [1907]

Wild bees, wasps and ants and other stinging insects./London, Routledge, [1907]./1 leaf, xiii, 144 p. illus. by Constance A. Saunders. 4 col. pl. incl. front./Bees pp. 1–87./A B G S . M

EDWARDES, E. Tickner, F.E.S. 439 1907

The bee-master of Warrilow./London, Pall Mall, 1907./64 p. 10 pl. front./ A G S . M W

2nd ed. new & enl./London, Methuen, 1920./219 p. 11 pl. front./B S
3rd ed./London, Methuen, 1921./219 p. 12 pl.

EDWARDES, [E.] Tickner 440 1908

The lore of the honey-bee./London, Methuen, 1908./xxiii, 281 p. 23 pl. front./B

2nd ed./London, Methuen, 1908./xxiii, 281 p. 23 pl. front./B G . M W
Numerous so-called eds. to 20th ed. 1949. B has 1st, 4th, 6th, 8th, 10th, 11th, 16th.

The photographs were omitted from the 4th ed. and are known to be absent from the 6th, 7th, 8th, 9th, 11th, 16th (Gateway Library Series), but present in the 10th.

Hodgson 339.

Pub. in U.S.A. 1911./Johansson 160.

Walker p. 20: 'For the general reader, the best and most informing work on the hive-bee hitherto published. Somewhat in the style of Maeterlinck's *La vie des abeilles*, but more accurate.'

The author's widow presented the original manuscript and photographs to B.

BALDWIN, S. J. 441 [1908]
The S.J.B. hints on bee-keeping./St. Albans (Herts.), Cartmell, [1908]./32 p. illus./A G

COWAN, T[homas] W[illiam], F.L.S., F.G.S., F.R.M.S., F.E.S., etc. 442 1908
Wax-craft: all about beeswax, its history, production, adulteration and commercial value./London, Sampson Low, Marston & British Bee Journal, 1908./172 p. illus. xvii pl. incl. front./A B G S . M W

In the proem Cowan states that nearly twenty years earlier he had translated from the German a small pamphlet of 19 pages on beeswax by M. J. Dennler (402).

B has translations of *Wax-craft* in French (1911) and Russian (1912).

CRIBB, F. J. 443 [1908]
Bee-keeping that pays./London, Cable Printing and Pub., [1908]./44 p. 6 plates./G

Anr. ed./[1910]./44 p. 6 pl./G S . M

HOPE, A. 444 1908
Honey and health./Glasgow, Maclehose, 1908./44 p./A S . M

TINSLEY, Joseph 445 [1908?]
The preparation of honey and wax for the show bench./Stafford, Chronicle, [1908?]./19 p./B

2nd rev. ed./[1912?]./43 p. illus./B

BENSUSAN, S. L. 446 1909
The children's story of the bee./London, Mills & Boon, 1909./viii, 248 p. 8 pl. front. by Carton Moore Park./A B G S . M

Hodgson 337.

GEARY, H[enry S.] 447 [1909]
Bees for profit and pleasure: a handbook dealing with bees and bee-keeping in all its various phases; with hints on the uses of honey, and a beekeeper's calendar. London, Collingridge, [1909]. 114 p. illus. incl. front./A B G S . M

Anr. ed. [1910]/A B

Profitable Farm and Garden Handbooks.

WHITE, C. N. 448 [1909]
Bees: a practical handbook./London, Vinton, [1909]./74 p. illus./A

Vinton's Country Series.

COWAN, T[homas] W[illiam], F.L.S., F.G.S., F.R.M.S. 449 [1910]
The queen bee (Apis mellifica): a popular illustrated record of British bee life, with dissectible model embracing both external and internal parts./London, British Bee Journal, [1910]./16 p. illus. 1 col. pl. 31 cm./B G S

A cut-out in the same style as 427, but not identical.

DIGBY, Sir Kenelm 1910
The closet of Sir Kenelm Digby, Knight, opened (*see* 42)

ABBOTT, S. Shapland 450 1911
The bee-keeper's companion./London, Mills & Boon, 1911./121 p. illus./BG

FABRE, J. H[enri] 451 1911
Social life in the insect world./London, Fisher Unwin, 1911./viii, 327 p. illus./*A bee-hunter*: the *Philanthus aviporus* pp. 150–178./A

Anr. ed./Harmondsworth (Middx.), Penguin, 1937./x, 11–245 p. XV pl./*A bee-hunter*: the *Philanthus aviporus* pp. 117–139./B

Translated by Bernard Miall from *Souvenirs entomologiques.*
The publication of Fabre's *Souvenirs* began in 1879. There were 10 volumes in all. For Vol. 1 translated into English see 421; see also 468 and 469.

GEARY, Henry [S.], F.E.S. 452 1911
Profitable bee-keeping for small-holders and others./London, Pearson, 1911./124 p. illus. 7 pl. front./A B G S . M

2nd ed./London, Pearson, 1917./124 p. illus. 7 pl. front./B
3rd ed./London, Pearson, 1918./124 p. illus. 7 pl. front./B
4th ed./London, Pearson, 1919./124 p. illus. 7 pl. front./B . M
New & rev. ed./London, Pearson, 1923./126 p. illus. 7 pl. front./B . M

The Smallholder Library No. 1.

AVERY, G. W. 453 1912
Bee keeping for beginners./Edinburgh, Edinburgh & East of Scotland College of Agriculture, 1912./31 p. illus./S . M

BANCKS, [Charles] Gerard W[instanley] 454 [1912]
The harvest of the hive./London, Unwin, [1912]./13 p./A G . M

BRITISH BEE-KEEPERS' ASSOCIATION 455 1912
Catalogue of the library of the B.B.K.A./London, British Bee-keepers' Association, 1912./29 p. 22 × 14 cm./S . M
Compiled by H. J. O. Walker; supersedes 362.

GRAHAM-SMITH, G. S. & others 456 1912 and 1913
Report on the Isle of Wight bee disease./London, H.M. Stationery Office, 1912./143 p. illus. VI pl. bibliog./B S
Further report on the Isle of Wight bee disease./London, H.M. Stationery Office, 1913./47 p. bibliog./B S
Supplements 8 & 10 of *Journal of the Board of Agriculture.*
Often catalogued under Board of Agriculture. B has one volume comprising the above reports and *Further report on a disease of bees in the Isle of Wight* by Walter Malden, *Journal of the Board of Agriculture* 15(11): 809–825 (1909).

HAWKS, Ellison 457 [1912]
Bees shown to the children./London & Edinburgh, Jack, [1912]./xii, 120 p. XXXVIII pl. front./A B G S . M
Shown to the Children Series No. 11.
Hodgson 332.
Pub. in U.S.A. under the title *Bees* 1912./Johansson 218.

HERROD, William, F.E.S.
(afterwards Herrod-Hempsall) 458 1912
Producing, preparing, exhibiting and judging bee produce./London, British Bee Journal, 1912./4 leaves, 168 p. illus. front./B G S . M
2nd rev. ed./London, B.B.J., 1948./viii, 219 p. illus./A

SLADEN, F. W. L., F.E.S. 459 1912
The humble-bee, its life-history and how to domesticate it: with descriptions of all the British species of *Bombus* and *Psithyrus.*/London, MacMillan, 1912/1 leaf, xiii, 283 p. illus. V col. pl. front./B G
The classical book on bumble bees. *Bee World* 41 : 8–9 (1960) gives a note about pencil amendments made by Sladen before he was drowned in Canada in 1922.

SNELGROVE, L. E. 460 1912
A method of requeening./Bristol, [Author], 1912./28 p. 14 cm./B

DUNCAN, F. Martin, F.R.M.S., F.R.P.S.
& DUNCAN, L. T. 460a [1913]
Bees, wasps and ants./London, Frowde, Hodder & Stoughton, [1913]./vi, 7–87 p. illus./Bees occupy 59 p. with 1 photographic and 2 col. pls. also illus. in text/A B
Often reprinted 1917–1936./A B S
For children./Hodgson 331.
Sometimes bound with 2 other works under the title *Insect builders and spinners.*

EDWARDES, [E.] Tickner 461 1913
The honey-star./London, Hutchinson, 1913./viii, 344 p./A B
Novel involving beekeepers.

LATTER, Oswald H., M.A., F.E.S. 462 1913
Bees and wasps./Cambridge, University Press, 1913./vi, 1 leaf, 132 p. illus. bibliog./*Apis mellifica* pp. 95–108./A B G S . W
Cambridge Manuals of Science and Literature.

[PURVIS, W.] 463 [1913]
Lloyd's penny bee book, by a member of the British Bee-keepers' Association./London, Lloyd's Weekly News, [1913]./64 p. illus./B . M
Walker (p. 58) describes it as 'A wonderfully efficient little book.'

MACDOUGALL, R. Stewart, M.A. 464 [1914]
Bee diseases and bee enemies./[Edinburgh?], Highland and Agricultural Society of Scotland, [1914]./20 p. illus. bibliog.

MORLEY, W. S. 465 1914
Bee-keeping for profit./London, Cassell, 1914./4 leaves, 124 p. 7 pl. front./A B S . W
Anr. ed./1919.

ROYDS, Thomas Fletcher, M.A., B.D. 466 1914
The beasts, birds and bees of Virgil: a naturalist's handbook to the Georgics./Oxford, Blackwell, 1914./*Bees* pp. 57–107./A B S
2nd rev. ed./Oxford, Blackwell, 1918./A B G

STOCKLEY, C. 467 1914
Wild honey./New York & London, Constable, 1914./340 p./A
Anr. ed./London, Constable, 1919./340 p./A
Anr. ed./London, John Long, 1925./340 p./A
Fiction. Misleading on description of bee behaviour.

FABRE, J. Henri 468 1914
The mason bees./London, Hodder & Stoughton, 1914./viii, 324 p./B S
Translated by Alexander Teixeira de Mattos from *Souvenirs entomologiques.* (*See* 451)

FABRE, J. Henri 469 1915
Bramble-bees and others./London, Hodder & Stoughton, 1915./viii, 476 p./B S
Translated by Alexander Teixeira de Mattos from *Souvenirs entomologiques.* (*See* 451)
Neither 468 nor 469 deals with honeybees.

NELSON, James Allen, Ph.D. 470 1915
The embryology of the honey bee./Princeton (U.S.A.), Princeton University, also London, Humphrey Milford, Oxford University Press, 1915./3 leaves, 282 p. 1 leaf, illus. 6 pl. bibliog./B
Johansson 371.
The classical work on the subject.

HERROD-HEMPSALL, William, F.E.S. (formerly Herrod) 471 [1915]
Bee-keeping simplified for the cottager and smallholder./London, Simpkin Marshall Hamilton & Kent, [1915]./3 leaves, 48 p. illus./A B G S . M
2nd ed./[1919]./A G . M
3rd ed./[1922]./A B . M
4th ed./[1929]./A

PHILLIPS, Everett Franklin, Ph.D. 472 1915
Beekeeping: a discussion of the life of the honey bee and of the production of honey./London, Macmillan, 1915./xxii, 1 leaf, 457 p. illus. 7 pl. front./A B . W
Anr. ed./1922.
Anr. ed./1937.
Anr. ed./1949./B U.S.A. only
Rural Science Series.
Published in U.S.A. 1915./Johansson 418.

SCOTTISH BEEKEEPERS' ASSOCIATION 473 1915

Library bye laws and condensed catalogue./1915./4 p./(B) S

Anr. ed./1921./8 p./A (B) S

Supplements 1924, 1930./A (B) S

Catalogue of the Moir Library. Scottish Beekeeper 9(10): 147–171 (1933)./B

Catalogue of the Moir Library./Stirling, Scottish Beekeepers' Association, 1950./193 p./A B S

Supplement 1950–1962./[Edinburgh], Scottish Beekeepers' Association, [1965]./iii, 97 p./B S

The 1950 *Catalogue* and its *Supplement* are valuable works of reference and have been indispensable in compiling the present *Bibliography*.

HERROD-HEMPSALL, W[illiam], F.E.S. 474 1916

Bees and their management./Chapter I of Vol. VI of *Live stock of the farm* ed. by C. Bryner Jones./London, Gresham, 1915–16. 6 vol. illus./G

This Vol., published in 1916, has 199 pp. and deals with bees, goats, dogs, ferrets, asses and mules. The bee chapter consists of 63 pp. with 13 figs. in the text and 2 photographic plates. This little-known work by Herrod-Hempsall is one of his earliest (*see* W. E. Saint *B.B.J.* 93 : 231 (1965)). There is a copy in the Library of the Ministry of Agriculture, but it is stored at Croydon.

ANDERSON, John, M.A., B.Sc.,
& RENNIE, John, D.Sc., F.R.S.E. 475 1916

Observations and experiments bearing on 'Isle of Wight' disease in hive bees. Reprint from *Proceedings of the Royal Physical Society of Edinburgh* 20(1): 23–61. 1 pl./[Edinburgh, Grant & Son], 1916./B

PELLETT, Frank C. 476 [c. 1916]

Productive bee-keeping: modern methods of production and marketing of honey./London, Lippincott, [c. 1916]./1 leaf, xiv, 302 p. illus. col. front./A B S . W

2nd ed. rev./London, Lippincott, [1918]./1 leaf, xiv, 302 p. illus. col. front./B S . W

3rd ed. rev./London, Lippincott, [1923]./1 leaf, xiv, 302 p. illus. col. front./A B . W

4th ed. rev./London, Lippincott, [1928]./xiv, ixa, 302 p. illus. col. front. bibliog./B

Lippincott's Farm Manuals.

RAYMENT, Tarlton 477 [1916]
Money in bees in Australasia: a practical treatise on the profitable management of the honey bee in Australasia./Melbourne (Australia) & London, Whitcombe & Tombs, [1916]./xvi, 293 p. illus. incl. front./A B

2nd ed. Melbourne & London, Whitcombe & Tombs, [1925]./xvi, 280 p. illus. incl. front. 1 col. pl./B

[GRIEVE, Mrs. M.] 478 [1917]
Bee-plants and their honey. [Chalfont St. Peter (Bucks.), Author, 1917.]/15 p./B P . W

The information is similar to that in a series of articles in *Bee World* 1922–23. The author's *A modern herbal*, 915 p., 96 pl., was edited by C. F. Leyel and published in 1931, reprinted 1974./B

McCALEB, Walter Flavius 479 1917
Happy, the life of a bee./London, Harper, 1917./120 p. illus./A . W

For children./Hodgson 330.
Pub. in U.S.A. 1917./Johansson 315.

HERROD-HEMPSALL, W[illiam], F.E.S. 480 1918
Bee-keeping in war-time./London, Country Life & George Newnes, 1918./32 p./A G P S . M W

2nd impression as *Beekeeping.*/London, Country Life, 1920./32 p./A S

Cover-title: A practical booklet on the best and most up-to-date methods of bee-keeping.

SMITH, Madge S. 481 1918
The bee book for young people./London, Stead, 1918./48 p. illus. by Brinsley le Fanu./Books for the Bairns No. 262. The Bairns Magazine for April 1918 (ed. by Estelle W. Stead) 16 p. follows at the end. The two together originally cost twopence as part of a monthly series.

TALBOT, Frederick A. 482 [1918]
Britain's greatest bee hunter: J. C. Bee Mason./From 'World's Work', [1918]./19 p. illus./B S

AUSTRALIAN IMPERIAL FORCE EDUCATION SERVICE 483 1919
Bee-keeping./London, Education Service A.I.F., 1919./27 p./B

A.I.F. Land Book No. 13.
Hodgson 324.

FABRE, J. Henri 484 [c. 1919]
The story book of science./London, Hodder & Stoughton, [c. 1919]./vi, 299 p./Chapter XXXIV pp. 278–299 is about honeybees.
For children.

LOVELL, J. H. 485 1919
The flower and the bee: plant life and pollination./London, Constable, 1919./xvii, 286 p. illus. front./A B G S
Hodgson 325.
Pub. in U.S.A. 1918./Johansson 308.

PELLETT, Frank C. 486 1919
Beginner's bee book./Philadelphia (U.S.A.) & London, Lippincott, 1919./179 p. illus. front./A B G S . W
Hodgson 328.
Johansson 405.

SMITH, S. H. 487 [1919]
Bee-keeping on modern principles: 'Let the bees tell you.'/Cambridge, Author, [1919]./48 p. illus./B
2nd ed./1919.
3rd ed./1920./S

CLARK, H. 488 [1920]
The lure of the hive: an interesting account of the life, habits, and work of the honey bee./Leicester, Stevens, [1920]./22 p./A

GEARY, Henry [S.] 489 1920
The beekeeper's vade-mecum./London, Stanley Paul, 1920./202 p. illus. incl. front. 15 × 7 cm./A B G S . W

MAETERLINCK, Maurice 490 [1920]
The children's life of the bee, selected and arranged by Alfred Sutro and Herschel Williams./London, Allen & Unwin, [1920]./191 p. illus. by E. J. Detmold./A
Anr. ed./London & Toronto, Dent, 1934./191 p./A
This ed. also has *Queen Bomba of the honey pots* and *Teddy Bear's bee tree* (a chapter from *Babes of the wild* by C. G. D. Roberts).
Anr. ed./London, Dent, 1949./191 p./B
King's Treasury of Literature.
For children./Hodgson 326.
Pub. in U.S.A. 1919./Johansson 323.
See also 422, 511.

OLDFIELD, J[osiah] 491 [1920]
Honey for health. 5th ed./Doddington (Kent), Fruitarian Society, [1920]./16 p. 15 cm./B
Best Food Series No. 10.

Anr. ed./Birmingham, [n.d.]./S
Anr. ed./London, Order of the Golden Age, [1952]./16 p./A
The Living Library.

Dr. Josiah Oldfield lived to be 100 at a time when few people did so. He ascribed his long life to his eating habits, and wrote many pamphlets on the subject such as *Eat right and live long*, *Eat and be happy*, *Fasting for health and life*, *Tuberculosis or Flesh-eating a cause of consumption*, *The voice of Nature or What man should eat*. He also wrote *The evils of butchery*, published by the Humanitarian League and *Hanging for murder* published by the Society for the Abolition of Capital Punishment.

FABRE, J. Henri 492 [1921]
Insect adventures./London, Hodder & Stoughton, [1921]./xii, 308 p./
Chapters III–VI pp. 24–85 deal with non-*Apis* bees.

Selections from Alexander Teixeira de Mattos's translation of *Souvenirs entomologiques* re-told for young people by Louise Seymour Hasbrouck. *See* 468, 469.
For children.

MACE, Herbert 493 [1921]
A book about the bee./London, Hutchinson, [1921]./x, 138 p., 14 pl. front./A B G S . W

Anr. ed./[1927].
Anr. ed./1930.

Hodgson 321, but not specifically for children.

ROLLE, Richard 1921
English writings. (*See ms* 10)

BOARD OF AGRICULTURE 494 1922
Collected leaflets on bee keeping./London, H.M. Stationery Office, 1922./1 leaf, 46 p. illus. 6 pl./B

2nd ed./1924./48 p. illus. 5 pl./B
For later eds. see 722.

BONSELS. W[aldemar] 495 [1922]
Maya: the adventures of a little bee. / London, Hutchinson, [1922]. / 259 p. illus. / A S

Anr. ed. / London, Hutchinson, [1934]. / 252 p. illus. / B
Anr. ed. / London, Hutchinson, [1946]. / 252 p. illus. / A

Hutchinson's Books for Young People. / Hodgson 322.
Pub. in U.S.A. 1922. / Johansson 64.
Translated by C. Remfry-Kidd from *Die Biene Maja und ihre Abenteuer.* / Frankfurt a. M., Rütten u. Loening, 1920.

DOUGLASS, Benjamin Wallace 496 1922
Every step in beekeeping. / London, Crosby Lockwood, 1922. / 6 leaves, 177 p., 1 leaf, 30 pl. front. / A B S

Anr. ed. / London, 1923. / illus. / S

Reissue of book pub. in U.S.A. 1921, Johansson 150.

497 [1922]
Paralysis, nosema, acarine, diseases of bees. / Sheffield, Newton Chambers, [n.d.] / 24 p. illus. / B
See 523. Recommends Izal for treatment.

PENNY, Henry S. 498 [1922]
Bee-keeping: a guide for beginners. / Honiton (Devon), Devon Bee-Keepers' Association, [1922]. / 29 p., 1 leaf. illus. / B

RENNIE, John, F.R.S.E., F.A.C. 499 1922
The crucial problem of British beekeeping. / Benson (Oxon.), Apis Club, 1922. / 8 p. / A B S . M

Reprinted from *Bee World* 4(3) : 72–74 (1922).

SNELGROVE, L. E. 499a 1922
Roman bee lore. A typed manuscript of 269 pages, complete as for printing, with preface, acknowledgements, illustrations and bibliography. / B

The manuscript was presented by the author's daughter, the late Miss I. Snelgrove; it is not known why the book was never published.

STURGES, A. M. 500 [1922]
Rational bee-keeping and the prevention of acarine disease. / Hartford (Ches.), Author, [1922]. / 13 p., 1 leaf. / B S

BETTS, Annie D., B.Sc. 501 1923

Practical bee anatomy: with notes on the embryology, metamorphoses and physiology of the honey bee./Benson (Oxon.), Apis Club, 1923./88 p. XII pl. bibliog./A B G . M W

Apis Club Library Vol. 1 (no subsequent Vol. known).

Miss Betts was one of the leaders of the Apis Club movement (*see* 561, Chapter 2). Apart from one other book (see 545, 565), most of her many writings on bees were published in *The Bee World*, which she edited for the Apis Club from 1929 until she retired at the end of 1949 and Dr. Eva Crane succeeded her. Largely through Miss Betts, when the Apis Club was wound up in 1952, the Apis Club Library was passed to the Bee Research Association, together with the many volumes of her own meticulous notebooks. Miss Betts had been engaged in aeronautical research during the first World War.

BROWN, Herbert 502 1923

A bee melody./London & New York, Melrose, 1923./v, 274 p./A B S . M

A somewhat fanciful book, with interesting quotations and references. Johansson 70.

EDWARDES, [E.] Tickner 503 1923

Bee-keeping for all: a manual of honey-craft./London, Methuen, 1923./viii, 136 p. illus. 3 pl. front./A B G S . W

2nd ed./1925.
3rd ed. rev./London, Methuen, 1929./xii, 138 p. illus. 3 pl. front./B
4th ed. rev./London, Methuen, 1934./xii, 138 p. illus. 3 pl. front./B
5th ed. rev. & amplified./London, Methuen, 1939./xii, 221 p. illus. 3 pl. front./B S
6th ed./London, Methuen, 1941./xii, 221 p. illus. 3 pl. front./B S
7th ed./1943.
8th ed./London, Methuen, 1945./xii, 205 p. illus. 3 pl. front./B
9th ed./London, Methuen, 1947./xii, 205 p. illus. 3 pl. front./B

EVRARD, Eugène 504 1923

The mystery of the hive./London, Methuen, 1923./3 leaves, 369 p./A B S . W

Translated from the French by Bernard Miall.

MACE, Herbert, F.E.S. 505 [1923]

Adventures among bees./London, Hutchinson, [1923]./144 p. 16 pl./A S

Anr. ed. London, Hutchinson, [1924]./144 p. 17 pl. front./B
Anr. ed./1927.
Anr. ed./1930./G

RENNIE, John, D.Sc. 506 1923

Acarine disease explained./[Benson (Oxon), Apis Club, 1923.]/50 p. illus./A B S . M

Revised, reprinted and enlarged from *Notes on acarine disease*, *Bee World*: 2(12): 144–145 (1921), and monthly articles to 3(12): 285–287 (1922).

STURGES, Arthur M., B.Sc. 507 1923

The rational system of bee-keeping: for the prevention of disease and avoidance of swarming./[Hartford (Ches.), Author], 1923./60 p. illus./ A G

2nd ed. [Hartford (Ches.), Author], 1923./60 p. 1 leaf, illus. 12 cm./A B S

TAYLOR, E. H., Ltd. [Sir William Gavin] 508 [1923?]

Bees for beginners./Welwyn (Herts.), Taylor, [1923?]. xii, 118 p. 1 leaf, illus. front. grey cloth./A B S

Anr. ed./Welwyn (Herts.), Taylor, [1923?]. 6 leaves, 127 p. 2 leaves, illus. front. yellow cloth./A B G

The dates of both eds. are uncertain. The firm has a copy of the 1st ed. showing the alterations made in the later ed. Mr. Moir had written 1923 on the title page of S's copy of the first ed.; however, G's copy of the later ed. was purchased on 7 May 1923, so the first ed. was probably earlier than 1923. There must have been two printings of the later ed., because in some copies the drone and worker on p. 12 are incorrectly identified. The date 1925 in the A's catalogue might refer to a later printing, but their date 1930? for the 118 p. ed. must be wrong.
The frontispiece is a reproduction of White's (123).

ASH, E. C. 509 [1924]

Ants, bees and wasps: their lives, comedies and tragedies./London, Holden, [1924]./136 p. illus./A . W

Nature Lovers Library Vol. 23.

JACKSON, I. H., M.A. 510 1924

Bee-keeping for beginners./London, Blackie, 1924./112 p. illus. 8 pl. front./A B S . W

Reprinted with index [*c.* 1945]./114 p. illus. 8 pl. col. front./B

MAETERLINCK, M. 511 1924
The child's book of the bee, abridged (from *The children's life of the bee*) by D. M. Northcroft./London & Vienna, New Education Fellowship, 1924./31 p. illus./A

See 490.
For children.

RICHARDS, A. 512 1924
Some observations on queen rearing and introduction, with special reference to Monsieur Perret-Maisonneuve's recent inventions./London, British Bee Journal, 1924./8 p. illus./A S

STURGES, Arthur M., B.Sc. 513 1924
Practical bee-keeping./London, Cassell, 1924./xix, 308 p. illus. 16 pl. col. front., bibliog./A B G S . W

**BALFOUR-BROWNE, F.,
M.A., F.R.S.E., F.Z.S., F.E.S.** 514 1925
Concerning the habits of insects./Cambridge, University Press, 1925./x, 169 p. IX pl./Lecture II *The habits of bees and wasps* pp. 29–57./S

Based on lectures 'adapted to a juvenile auditory' at the Royal Institution in 1924.

EDWARDES, [E.] Tickner 515 1925
Beekeeping do's and dont's. London, Methuen, 1925./viii, 56 p./A B G S

FLOWER, A. B., F.E.S. 516 1925
Bee-keeping up to date./London, Cassell, 1925./x, 110 p., illus. 7 pl. front./A B S

2nd ed./1930.
3rd ed. rev./1934.
4th ed./1936.
5th ed./1939.
6th ed. rev./London, Cassell, 1940./x, 110 p. illus.
7th ed./1941./110 p. illus.
8th ed./London, Cassell, 1942./x, 119 p. illus. 8 pl./B S
6th ed. [*sic*] rev./1952./119 p. illus./A

Cassells Pet and Livestock Series.
Herrod-Hempsall writes about the author in his Foreword: 'Not only does she manage a home apiary of fair size, but also out-apiaries—one of the most difficult problems in beekeeping'.

HALL, Charles A. 517 1925
Bees, wasps and ants./London, Black, 1925./viii, 87 p. 12 pl./A S . W
Peeps at Nature Series for children. Hodgson 318.

HAYES, George 518 1925
Nectar producing plants and their pollen: with directions for the separation, examination and photographing of pollen grains from honey. Also drawings of pollen grains from foreign honeys./[London], British Bee Journal, 1925./110 p. illus./A B G S
An early attempt at such drawings.

HOPKINS, Isaac & MULVANY, T. J. 519 [1925]
Practical bee-keeping, being the sixth edition of The Australasian bee manual./Christchurch & London, Whitcombe & Tombs, [1925]./xvii, 288 p. illus./A B
Anr. ed./London, 1930./283 p./G S
Whitcombe's Practical Handbooks.
The Australasian bee manual was published in New Zealand as *The New Zealand Bee Manual* 1881, 2nd ed. 1882; the 3rd ed. was *The illustrated Australasian bee manual*, 1886.

MACE, Herbert, F.E.S. 520 [1925]
Some other bees, butterflies and moths./London, Hutchinson, [1925]./160 p. front illus. 14 pl./Bees pp. 11–40./A B S

SHARP, Dallas Lore 521 1925
The spirit of the hive: contemplations of a beekeeper./New York & London, Harper, 1925./6 leaves, 240 p. 2 leaves./A S
Johansson 528.

STRATTON-PORTER, Gene 522 1925
The keeper of the bees./London, Hutchinson, 1925./515 p./A
Anr. ed./London, Hutchinson, [1926]./312 p./A
Anr. ed./London, Universal Library, [1929]./315 p./B
Anr. ed./Bath (Somerset), Chivers, 1969./312 p./A
Fiction. Pub. in U.S.A. 1925./Johansson 433.

EDWARDES, [E.] Tickner, F.E.S.,
& STURGES, Arthur M., B.Sc. 523 [1926]
Apiary hygiene./Sheffield, Newton Chambers, [1926]./27 p. illus./B
Issued by the proprietors of Izal disinfectant to promote its use in the apiary to combat disease.
See 497.

(*cancelled*) 524–530

1927 to 1976

Author bibliography for the last fifty years

A., J. 1928 & 1930
Directions for the making of colonies for bees. Reprint of 1683 ed. (*see* 58).

ABBOT, A. E. *editor & compiler* 531 1963
Honey as healer. / London, Emerson Press, 1963. / 35 p. bibliog. / B
2nd, 3rd eds. (*see* 786).

ABBOTT, C. P. 532 1947
Queen breeding for amateurs. / Petts Wood (Kent), Bee Craft, 1947. / 48 p. 8 pl. by author. / B
New and rev. ed. / Petts Wood (Kent), Bee Craft, 1951. / 60 p. 10 pl. by author. / B

ABBOTT, May 533 1965
Me and the bee: the battles of a bumbling beekeeper. / London, Hammond Hammond, 1965. / 189 p. illus. / B

ACTON, Bryan & DUNCAN, Peter 534 1965
Making mead—a complete guide to the making of sweet and dry mead, melomel, metheglin, hippocras, pyment and cyser. / Andover (Hants.), Amateur Winemaker, 1965. / 60 p. 8 pl. 18 cm. / B

ADAM, Brother 535 1975
Bee-keeping at Buckfast Abbey: with a section on meadmaking. / Geddington (Northants.), British Bee Publications, 1975. / 86 p. 14 pl. / B
Based in part on *Meine Betriebsweise* 1971.

ADRIAN, Mary *pseud.* **(VENN, Mary Eleanor)** 536 1964
Honeybee tells honeybee. / Kingswood (Surrey), World's Work, 1964. / 51 p. illus. by Barbara Latham, front. / B
Life-Cycle Stories.
For children. / Hodgson 232.
Originally pub. in U.S.A. 1952 as *Honeybee*, as a Holiday House Easy Science Book. / Johansson 579.

ÅKERBERG, Erik & CRANE, Eva *editors* 537 1966
Proceedings of the Second International Symposium on Pollination held in London, July 1964./London, Bee Research Association, 1966./216 p. illus. bibliog./B
Supplement to *Bee World* Vol. 47, 1966.

ALFORD, D. V. 538 1973
Bumblebee Distribution Maps Scheme. Guide to the British species./London, Bee Research Association, 1973./48 p. illus./B
Reprinted from articles in *Entomologist's Gazette.*

ALFORD, D. V. *editor* 539 1973
for Bee Research Association
Bumblebees: provisional atlas of the insects of the British Isles. Part 3. Hymenoptera—Apidae (*Bombus : Psithyrus*)./Abbots Ripton (Huntingdon), Biological Records Centre, 1973./31 p. incl. 27 maps. 27 cm./B
Supersedes 558.
John Heath is General editor of the *Atlas*, which is linked with the European Invertebrate Survey.

ALFORD, D. V. 540 1975
Bumblebees./London, Davis-Poynter, 1975./xii, 352 p. illus. 56 pl. (16 col.), 27 maps. bibliog. 25 cm./B

ALLEN, A. L. Sandeman 541 1952
Bee-keeping with twenty hives./Petts Wood (Kent), Bee Craft, 1952./126 p. illus. by N. Nettleton, 2 pl./B

ANDERSON, J. 542 1938
Bees, honey and beekeeping./[Glasgow], Scottish Beekeepers' Association, 1938./36 p. 2 pl. incl. in pagination./B

ANDREWES, Sir Christopher 543 1969
The lives of wasps and bees./London, Chatto & Windus, 1969./204 p. 16 pl. bibliog./B

ANDREWS, S. W. 544 1971
All about mead./London, Mills & Boon, 1971./96 p./B

APIS CLUB 545 [1934]
The diseases of bees: their signs, causes and treatment./[Camberley (Surrey), The Apis Club, 1934]/2 leaves, 49 p. 8 pl./B
See 565.

ARMITT, J. Harold 546 [1952]
Bee-keeping for recreation and profit: relating craftsmanship to the bee colony's natural development and instinctive behaviour./Birmingham, Ladbrook, [1952]./98 p. illus. 6 pl./B

ASTON, Oliver 547 1975
Bee and wasp./Oxford, Blackwell, 1975./31 p. illus./A
Adapted from *Rosenbi og snyltehveps* by Mogens Jansen, pub. in Denmark, 1974.
For children.

AUBREY, John 1969
Natural history of Wiltshire . . . Reprint of 1847 ed. (*see* 63)

BAILEY, Leslie 548 1963
Infectious diseases of the honey-bee./London, Land Books, 1963./176 p. illus. by J. P. Spradbery, 8 pl. front. bibliog./B

BAKER, C. T. G. 549 [1948]
Understanding the honey bee: notes on the organic nature of the bee colony, and its relation to plant and cosmos./[Clent (Worcs.), Anthroposophical Agricultural Foundation, 1948]./32 p. illus. 6 pl./B

BANNER, Angela 550 [1951]
Ant and bee./London, Edmund Ward, [1951]./109 p. col. illus. by Bryan Ward, 10 × 12 cm./B
For children, but no information on bees./Hodgson 236.
This Book 1 is followed by Books 2–12, see Hodgson 107, 138–140, 146, 151, 157, 184, 197, also Book 0 in which Ant and Bee meet.

BARRATT, Gilbert 551 [1931 ?]
Whole-time bee-keeping. Four addresses./Salisbury (Wilts.), Wilts., Hants. & Dorset B.K.A., [1931 ?]./16 p./B
Originally broadcast by the B.B.C.

BEBBINGTON, Nicolas 552 [1945?]
Visiting Mr. Bee./London, Blandford Press, [1945?]./69 p. col. illus. by R. M. Turvey./B

For children./Hodgson 267.

BECK, Bodog F. 553 1935
Bee venom therapy: bee venom, its nature and its effect on arthritic and rheumatoid conditions./New York & London, Appleton Century, 1935./xii, 2 leaves, 238 p. bibliog./B

Johansson 52.

BECK, Bodog Felix & SMEDLEY, Dorée 554 1947
Honey and your health: a nutrimental, medicinal and historical commentary. London, Museum Press, 1947./231 p. illus. 12 pl./A

Anr. ed./London, Bantam Books, 1971./xiv, 207 p./B

Originally pub. in U.S.A. 1938 as *Honey and health* by B. F. Beck.
Rev. ed. 1944 *Honey and your health* by B. F. Beck & D. Smedley.
Johansson 53.

BEE CRAFT 555 [1940?]
Apiary record book and guide./Petts Wood (Kent), Bee Craft, [1940?]. 71 p. 12 cm./B

BEE RESEARCH ASSOCIATION 556 1966
Bee research directory: a world guide to bee research workers and institutions./London, Bee Research Association, 1966./142 p. 1 leaf./B

Replaced 597.

BEE RESEARCH ASSOCIATION 557 1968
English alphabetical subject index to Universal Decimal Classification numbers used by the Bee Research Association in *Apicultural Abstracts* and in Library subject indexes./London, Bee Research Association, 1968. 117 p. 26 cm./B

2nd ed./London, Bee Research Association, 1971./144 p. 22 cm./B

1st ed. known as EASI, 2nd ed. as EASI 2, 3rd ed. EASI 3 due 1979.
Addenda is 560 (*see also* 570).

BEE RESEARCH ASSOCIATION 558 1973
Bumble Bee Distribution Maps Scheme. Preliminary bumble bee atlas. London, Bee Research Association, 1973./31 p. 27 maps./B

Provisional ed. superseded by 539.

BEE RESEARCH ASSOCIATION 559 1973
World list of films on bees and beekeeping. / London, Bee Research Association, 1973. / vii, 67 p. 30 cm. / B
Provisional ed.

BEE RESEARCH ASSOCIATION 560 1974
Addenda to English alphabetical subject index to Universal Decimal Classification numbers used by the Bee Research Association in *Apicultural Abstracts* and in subject indexes, second enlarged edition 1971. / London, Bee Research Association, 1974. / 1 leaf, 52 p. / B
See 557.

BEE RESEARCH ASSOCIATION 561 1974
Bee Research Association, 1949–1974: a history of the first 25 years. London, Bee Research Association, 1974. / 200 p. illus. / B
Supplement to *Bee World* Vol. 55, 1974.

BELL, M. G. Kennedy 562 1932
The joys of bee-keeping. / London, Herbert Jenkins, 1932. / 191 p. illus. 4 pl. front. / B
Hodgson 307.

BENT, E. R. 563 1946
Swarm control survey. / Aldershot (Hants.), Gale & Polden, 1946. / 5 leaves, 87 p. col. front. / B

BENT, E. R. 564 1950
A bee-keeper's progress. / Aldershot (Hants.), Gale & Polden, 1950. / ix, 142 p. illus. 1 pl. front. bibliog. / B

BETTS, Annie D. 565 [1934]
The diseases of bees: their signs, causes and treatment. / Camberley (Surrey), Apis Club, [1934]. / 49 p. illus. 4 pl. / B
2nd ed. / Camberley (Surrey), Hickmott for the Author, 1951. / 69 p. illus. 4 pl. / B
1st ed. pub. without author's name (545).

BINDING, G. J. 566 1971
About pollen. / London, Thorsons, 1971. / 63 p. / B
'About' Series.

BONE, David 567 1952
The bees of Westerham./Beckenham (Kent), Author, 1952./3 leaves, 18 p./B
Verse.

BONE, David 568 1957
The bees of Swanland./Beckenham (Kent), Bee & Blackthorn Press, 1957./5 leaves, 158 p. 2 leaves, illus. by Arthur Coombes & the author, front./B
Verse.

BONSELS, W[aldemar] [1934] & [1946]
Maya: the adventures of a little bee. (eds. of 495)

BRETT, Walter *editor* 569 1940
Honey from your garden: a handy guide to bee-keeping./London, C. Arthur Pearson, 1940./54 p./B
Reprinted 1940, 1941, 1942.
War Food-growing Handbooks No. 3.

BRITISH STANDARDS INSTITUTION 570 1969
Universal Decimal Classification. UDC 636/639: animal breeding, animal produce, hunting, fishing./London, British Standards Institution, 1969./37 p. 29 cm./B
BS 1000 (636/639): 1969. Part of the *Universal Decimal Classification* English Full Edition, Fourth International Edition (FID Publication No. 179). Includes 638.1 Beekeeping; *see* 557, 560.

BROWN, R. H. 571 1975
One thousand years of Devon beekeeping./Devon, Devon Beekeepers Association, 1975./2 leaves, 65 p. illus./B

BROWNING, Gareth H. 572 1933
The children's story-book of bees./London, Burns Oates & Washbourne, 1933./viii, 133 p. 6 pl./B

BUCKLEY, Arabella Burton 1929
The fairyland of science. Reprint of 1903 ed. (*see* 344)

BUHET, Gil 573 1953
The honey siege./London, Jonathan Cape, 1953./253 p./B
Anr. ed./Harmondsworth (Middx.), Penguin Books, 1958./248 p./B
Fiction, translated by Geoffrey Sainsbury from *Le chevalier pierrot* 1952.

BUNN-RICHARDS, Joan 574 1962
Secrets of the bees./London, Ward Lock, 1962./20 p. col. illus. by Marianne Scheel, 25 cm./B
For children./Hodgson 165.

BUTLER, Colin G. 575 [1948]
An introduction to bee keeping./London, Daily Mail School-Aid, [1948]. 23 p. illus. by J. Yunge-Bateman, bibliog./B
Young Britain Educational Series.
For children./Hodgson 251.

BUTLER, C[olin] G. 576 1949
Bee-hives./London, H.M. Stationery Office, 1949./ii, 20 p. 4 pl./B
2nd ed./London, H.M. Stationery Office, 1952./iv, 20 p. illus. 4 pl./B
Ministry of Agriculture & Fisheries Bulletin No. 144 (*see* 729).

BUTLER, C[olin] G. 577 1949
The honeybee: an introduction to her sense-physiology and behaviour./ Oxford, Clarendon Press, 1949./viii, 139 p. VI pl. bibliog./B

BUTLER, Colin G. 578 1954
The world of the honey bee./London, Collins, 1954./xiv, 226 p. II col. pl. 40 pl. by author, incl. front. bibliog./B
Rev. ed./London, Collins, 1962./xiv, 226 p. II col. pl. 40 pl. by author incl. front. bibliog./B
Rev. ed./London, Collins, 1974./xiv, 226 p. II col. pl. 40 pl. by author incl. front. bibliog./B
New Naturalist Series No. 29.
Hodgson 214.

BUXTON, Gene 579 [1947]
The bees wedding./London, Quality Press, [1947]./63 p. 8 pl. by K. Hunter./A
For children./Hodgson 256.

BUZZARD, C. N. 580 [1946]
Bomba the bumble bee./London, Daily Mail School-Aid, [1946]./15 p. illus. by J. Yunge-Bateman./B
Young Britain Educational Series.
For children./Hodgson 261.

BUZZARD, C. N. 581 194.
Shining hours. / London, Collins, 1946. / 192 p. illus. by J. Yunge-Bateman 8 pl. / B

Reminiscences about bees and beekeeping in the south of France.

Anr. ed. / London, The Country Book Club, 1952. / 192 p. illus. by J. Yunge-Bateman, 8 pl. / B

CAMPBELL, D. J. & HENDERSON, G. P. 581a 1962
The Bee World: index to Volumes 1–30, 1919–1949. / London, Bee Research Association, 1962. / xx, 119 p. 28 cm. / B

Dedicated to the Editors of the journal:

1919–1926 Dr. Ahmed Z. Abushâdy
1926–1929 James B. Morgan
1929–1949 Annie D. Betts.

CAMPBELL, Peter 582 1969
Harry's bee. / London, Methuen, 1969. / 27 p. col. illus. 27 cm. / B

For children. / Hodgson 113.

CARTER, George A. [J.] 583 1945
Bees and honey: a guide to the better understanding of bees, their diseases, and the chemistry of bee-keeping. / Petts Wood (Kent), Smith, 1945.

2nd ed. / Petts Wood (Kent), Bee Craft, 1946. / viii, 9–114 p. / B
3rd ed. / Petts Wood (Kent), Bee Craft, 1948. / viii, 9–115 p. / B

CARTER, George A. [J.] 584 [1946]
The hive bee. / Worcester, Littlebury, [1946]. / 44 p. 12 pl. bibliog. / B

CARTLAND, Barbara 585 1970
The magic of honey. / London, Corgi Books, 1970. / 158 p. 14 cm. / B

CARTLAND, Barbara 586 1976
The magic of honey cookbook. / London, Corgi Books, 1976. / 110 p. / **B**

CHAPMAN, A. D'Arcy 587 1936
A honey-bee and her master. / Oxford, Blackwell, 1936. / 4 leaves, 239 p. illus. front. / B

Reprinted 1944.

For children. / Hodgson 295.

CHAUNCY, Nan 588 1967
Beekeeping./London, Oxford University Press, 1967./32 p. illus. by Jane Walker./B

Life in Australia Series.

For children./Hodgson 126.

CHESNAKOV, Alexis 589 [1950]
The book of bees./London, Owl Press, [1950]./48 p. illus. by Basil Jonzen, 18 × 25 cm./B

For children./Hodgson 243.

CLARK, K[enneth] K. 590 1951
Beekeeping./Harmondsworth (Middx.), Penguin Books, 1951./222 p. illus. 8 pl./B

Penguin Handbooks Series.

CLOUDSLEY-THOMPSON, J. L. 591 1974
Bees and wasps./London, Bodley Head, 1974./48 p. illus. by Joyce Bee./B

A Bodley Head New Biology.
For children.

COCKERELL, T. D. A. 591a 1937
African bees of the genera Ceratina, Halictus *and* Megachile./London, British Museum (Natural History), 1937./xvi, 254 p. illus./B

COOK, J. H. 592 1927
Honey and the bee./Birmingham, Pitman, 1927./29 p./A

Health from Food Library No. 12.

CORYN, Marjorie 593 1947
A swarm of bees./London, Hodder & Stoughton, 1947./384 p./B

Fiction, historical.

COTTON, William Charles 1970
My bee book. Reprint of 1842 ed. (*see* 238)

COUSTON, Robert 594 1972
Principles of practical beekeeping./Kilmarnock (Ayrshire), Scottish & Universal Newspapers, 1972./4 leaves, 101 p. illus. 24 cm./B

COWAN, T[homas] W[illiam] 595 [1928]
British Bee-keepers' Association jubilee: history of the Association representing fifty years of bee-keeping progress. / [London, British Bee Journal, 1928]. / 2 leaves, 145 p. illus. front. / B

Reprinted from *British Bee Journal* with a portrait.
For earlier books by T. W. Cowan, see Author Index.

CRANE, Eva *editor* 596 1951
Dictionary of beekeeping terms with allied scientific terms. Giving translations from and into English–French–German–Dutch with Latin index. / London, Bee Research Association, 1951. / 74 p. 1 leaf. / B

Vol. I of a series; see 599, 600, 601, 794.

CRANE, Eva *editor* 597 1953
World list of bee research workers. / London, Bee Research Association, 1953. / 39 p. 25 cm. / B

Supplement, 1954. / 8 p. / B
New ed. / London, Bee Research Association, 1958. / 72 p. map. / B
Replaced by 556.

CRANE, Eva & others 598 1955
Beekeeping research digest. / [London], Bee Research Association, 1955. / 32 p. / B

CRANE, Eva *editor* 599 1958
Dictionary of beekeeping terms with allied scientific terms. Vol. II. Giving translations from and into English–Italian–Spanish. / London, Bee Research Association, 1958. / 63 p. / B

CRANE, Eva *editor* 600 1964
Dictionary of beekeeping terms with allied scientific terms. Vol. III. Giving translations from and into English–French–German–Czech–Polish–Russian with Latin index. / Warsaw, Państwowe Wydawnictwo Rolnicze i Leśne [in association with the Bee Research Association, London], 1964. / x, 203 p. / B

Editors for Slavic languages A. Demianowicz & L. Bornus.

CRANE, Eva *editor* 601 1971
Dictionary of beekeeping terms with allied scientific terms. Vol. IV. Giving translations from and into English–Danish–Norwegian–Swedish. / London, Bee Research Association, 1971. / 108 p. / B

Editor for Scandinavian languages G. Kolisko.
Vol. I, II & III are respectively 596, 599 & 600; Vol.V–VIII follow.

CRANE, Eva *editor* 602 1972
Pollination of seed crops: a summary of research reported in *Apicultural Abstracts* 1959–1971./London, Bee Research Association, 1972./163 p./B
Consists of 583 entries from *Apicultural Abstracts*, arranged under crop and under pollinator.

CRANE, Eva *editor* 603 1975
Honey: a comprehensive survey./London, Heinemann in co-operation with the Bee Research Association, 1975./xvi, 608 p. illus. 30 pl. bibliog./B
Reprinted with corrections 1976./B

CRANE, Eva & TOWNSEND, Gordon F. 604 1976
Index to Apicultural Abstracts 1950–1972./[Folkestone (Kent)], Dawson for Bee Research Association, 1976./*22*, 1824 p. in 2 vol. 30 cm./B
Vol. 1 is *22*, pp. 1–708 (includes Author index).
Vol. 2 is pp. 709–1824 (Subject index).

CRANE, Eva *editor* 604a 1976
Apiculture in tropical climates. Full report of the First Conference on Apiculture in Tropical Climates./London, International Bee Research Association, 1976./x, 208 p. illus. bibliog./B

CROMPTON, John *pseud.*
(LAMBURN, John B. C.) 605 [1947]
The hive./Edinburgh & London, Blackwood, [1947]./3 leaves, 174 p. illus. by A. E. Bestall, col. front./B

CROSS, Rena 606 1974
The honey and yoghurt cookbook./London, W. Foulsham, 1974./73 p. 2 leaves./B

CROWDER, Dorothy E. 607 [1951]
The flying nation./London, Hutchinson, [1951]./156 p. illus. by Helen Haywood, 8 col. pl. front./B
Hutchinson's Books for Young People Series.
For children./Hodgson 238.

CUMMING, A. R. 608 [1942]
The northern beekeeper: handbook of the Inverness-shire Beekeepers' Association./Inverness, [for Inverness-shire Beekeepers' Association], [1942]./56 p./B
2nd ed./Inverness, [for Inverness-shire Beekeepers' Association], 1945./100 p. illus.

CUMMING, A. R. & LOGAN, Margaret 609 1950
Beekeeping: craft and hobby./Edinburgh & London, Oliver & Boyd, 1950./ x, 157 p. illus. XVI pl. incl. front./B

DADE, H. A. 610 1949
The laboratory diagnosis of honey-bee diseases./London, Williams & Norgate, 1949./19 p. illus. 4 pl./B
Monographs of Quekett Microscopical Club No. 4.

DADE, H. A. 611 1962
Anatomy and dissection of the honeybee./London, Bee Research Association, 1962./xi, 158 p. 1 leaf, illus. 20 pull-out pl. bibliog./B
Reprinted with 5 p. of Notes and corrections./London, International Bee Research Association, 1977./B

DAVIS, G. H. 612 [1976]
Down to the grass roots of beekeeping: the management of honey bees in the swarming season./Malvern (Worcs.), Author, [1976]./32 p. illus./B

DEANS, Alexander S. C. 613 1949
The bee keepers encyclopedia./Kingswood (Surrey), Elliot, 1949./190 p. illus. 14 pl. front. bibliog./B
Right Way Books.
2nd rev. ed./Kingswood (Surrey), Elliot, [n.d.]/B

DEANS, A[lexander] S. C. 614 1957
Survey of British honey sources./[London], Bee Research Association, 1957./20 p. tables. bibliog./B
Report on pollen analyses of 854 samples.
Reference BRA 142.

DEANS, A[lexander] S. C. 615 1962
Bees and beekeeping./Edinburgh & London, Oliver & Boyd, 1962./iv, 76 p. illus./B
Quest Library.
For children./Hodgson 152.

DEANS A[lexander] S. C. 616 1963
Beekeeping techniques./Edinburgh & London, Oliver & Boyd, 1963./xii, 1 leaf, 160 p. 9 pl. bibliog./B

DIGGES, Rev. J[oseph] G. 1928 to 1950
The practical bee guide. (6th–16th eds. of 428)

DINES, Arthur M. 617 1968
Honeybees from close up./London, Cassell, 1968./4 leaves, 114 p. illus. with photographs by Stephen Dalton and others, front. bibliog. 26 cm./B

DODD, William 618 [1974]
Cumberland Beekeepers' Association: a history compiled from official records./[Cumbria, Cumberland Beekeepers' Association, 1974]./59 p./B

DOERING, Harald 619 1962
A bee is born./London, Oak Tree Press, 1962./96 p. illus. by author./B
Reprinted 1968.
Sterling Nature Series.
Translated and adapted by Dale S. Cunningham from *Königreich im Bienenkorb* 1962.
For children./Hodgson 153.
Pub. in U.S.A. 1962./Johansson 145.

DUGAT, M. 620 1948
The skyscraper hive./London, Faber & Faber, 1948./xvi, 78 p. illus. 4 pl. bibliog./B
Translated by Norman C. Reeves from *La ruche gratte-ciel à plusieurs reines* 1946.

DUMMELOW, John 621 1973
The Wax Chandlers of London: a short history of the Worshipful Company of Wax Chandlers, London./London & Chichester, Phillimore, 1973./xvi, 1 leaf, 204 p. illus. 18 pl. col. front. bibliog. 25 cm./B

DUNNING, J. M. 622 1945
The key of the hive: the habits and products of the honey-bee./London, Kingsgate Press, 1945./128 p. 16 pl. front./B

DURUZ, R. M. & CRANE, Eva 623 1953
English bee boles./London, Bee Research Association, 1953./16 p. illus./B
National Beekeeping Museum Pamphlet No. 1; no others published.

ECKERT, John E[dward] 624 1960
& SHAW, Frank R[obert]
Beekeeping. / New York & London, Macmillan, 1960. / ix, 1 leaf, 536 p. illus. front. bibliog. / B

Successor to *Beekeeping* by E. F. Phillips. / Johansson 418.

(*cancelled*) 625

EDWARDES, E. Tickner 1933 to 1949
The lore of the honey-bee. (other eds of 440)

EDWARDES, E. Tickner 1929 to 1947
Bee-keeping for all. (3rd to 9th eds of 503)

ELKIN, Benjamin 626 1971
The wisest man in the world: a legend of ancient Israel. / London, World's Work, 1971. / 38 p. 1 leaf, col. illus. by Anita Lobel. 25 cm. / B

For children. / Hodgson 102.

ELLSWORTH, E. E. 627 [1941]
Buzzywing. / London & Glasgow, Blackie, [1941]. / 64 p. illus. incl. front. / B

Blackie's Easy to Read Books.

For children. / Hodgson 278.

EMPIRE MARKETING BOARD 628 1932
The demand for honey: report of an investigation by the statistics and intelligence branch of the Empire Marketing Board into the retail marketing of honey. / London, H.M. Stationery Office, 1932. / 68 p. (pp. 65–68 are advert.) 4 pl. / B

EVERETT, Eileen 629 1966
Bees. / Exeter (Devon), Wheaton, 1966. / 24 p. col. illus. / B

Read about it Series.
For children. / Hodgson 120.

EWALD, Carl 1930
The queen bee. (anr. ed. of 437)

FABRE, J. H[enri]
Social life in the insect world. (anr. ed. of 451) 1937

FAEGRI, K. & PIJL, L. van der 630 1966
The principles of pollination ecology./Oxford, Pergamon Press, 1966./ix, 248 p. illus. bibliog./B

2nd rev. ed./Oxford, Pergamon Press, 1971./xii, 291 p. illus. bibliog./B

FLOWER, A. B. 1930 to 1952
Bee-keeping up to date. (2nd to 9th eds of 516)

FOSTER, Virgil E. 631 1962
Close-up of a honey-bee./London, Gollancz, 1962./63 p. illus. with photographs by Martin Iger. 20 × 24 cm./B

For children./Hodgson 154

Pub. in U.S.A. 1960./Johansson 177.

FRANÇON, Julien 632 1939
The mind of the bees./London, Methuen, 1939./xi, 146 p. illus./B

2nd ed./London, Methuen, 1947./xi, 143 p. illus./G

Translated by H. Eltringham from *L'esprit des abeilles.*

FRASER, H. Malcolm 633 1931
Beekeeping in antiquity./London, University of London Press, 1931./xi, 157 p. 8 pl. front. bibliog./B

2nd ed./London, University of London Press, 1951./xi, 145 p. 8 pl. front. bibliog.

FRASER, H. Malcolm 634 1951
Anton Janscha on the swarming of bees: a critical survey./[Foxton (Herts.)], Apis Club, 1951./28 p. incl. front./B

FRASER, H. Malcolm 635 1958
History of beekeeping in Britain./London, Bee Research Association, 1958./106 p. 7 pl. front. bibliog./B

Reprinted from articles in *Bee World.*

FREE, John B. 635a 1970
Insect pollination of crops./London and New York, Academic Press, 1970./xi, 544 p. illus.

FREE, John B. & BUTLER, Colin G. 636 1959
Bumblebees: with two appendices by Ian H. H. Yarrow./London, Collins, 1959./xiv, 208 p. 1 leaf, illus. 24 pl. col. front. bibliog./B

Reprinted 1968.

New Naturalist Series No. 40

FRISCH, Karl von 637 1950

Bees: their vision, chemical senses and language./Ithaca (N.Y.), Cornell University Press & London, Oxford University Press, 1950./xiii, 1 leaf, 119 p. illus. bibliog./**B**

Reprinted 1951.

Anr. ed./London, Jonathan Cape, 1968./125 p. illus. 8 pl. bibliog./**B**
Rev. ed./Ithaca (N.Y.) & London, 1971./xviii, 157 p. illus. bibliog./**B**
Johansson 183.

FRISCH, Karl von 638 1954

The dancing bees: an account of the life and senses of the honey bee./London, Methuen, 1954./xiv, 183 p. illus. XXX pl./**B**

Readers Union ed./London, Methuen, 1955./xiv, 183 p. illus. XXX pl./
2nd rev. & enlarged ed./1966./xiv, 198 p. illus./**B**

Translated by Dora Ilse from *Aus dem Leben der Bienen*, 5th ed. 1953.
Revisions and additions translated by Norman Walker from 7th ed. 1964.
Hodgson 133.

FRISCH, Karl von 639 1967

The dance language and orientation of bees./London, Oxford University Press, 1967./xiv, 566 p. illus. bibliog. 25 cm./**B**

Translated by Leigh E. Chadwick from *Tanzsprache und Orientierung der Bienen* 1965.

Pub. in U.S.A. 1967./Johansson 185.

FURNESS, Clara A. *compiler* 640 1974

Mary Workman's honey recipes./Thulston (Derbys.), British Isles Bee Breeders' Association, 1974./[54] p./**B**

GALTON, Dorothy 641 1971

Survey of a thousand years of beekeeping in Russia./London, Bee Research Association, 1971./90 p. illus. bibliog./**B**

GAMBLE, Reginald 642 1939

Bee-keeping./Radlett (Herts.), National Federation of Young Farmers' Clubs, 1939./48 p. illus. with photographs by R. V. Roberts, bibliog./**B**

2nd ed./Radlett (Herts.), National Federation of Young Farmers' Clubs, 1940./48 p. illus. bibliog.
3rd ed./Radlett (Herts.), National Federation of Young Farmers' Clubs, 1942./52 p. (pp. 49–52 are advert.), illus. bibliog./**B**

4th ed./Radlett (Herts.), National Federation of Young Farmers' Clubs, 1943./48 p. illus. bibliog./B
8th ed./Radlett (Herts.), National Federation of Young Farmers' Clubs, 1947./48 p. illus. bibliog./B
10th ed./London, Evans Bros for the National Federation of Young Farmers' Clubs, 1950./48 p. illus. bibliog./B
11th ed./London, Evans Bros for the National Federation of Young Farmers' Clubs, 1957./48 p. illus. bibliog./B

Young Farmers' Club Booklet No. 2.
Hodgson 283.

GAYRE, G. R. 643 1948
Wassail! in mazers of mead: an account of mead, metheglin, sack and other ancient liquors, and of the mazer cups out of which they were drunk with some comment upon the drinking customs of our forebears./London, Phillimore, 1948./176 p. illus. front./B

GERARD, John 1975
The herball. (reprint of anr. ed. of 14)

GILMAN, A. 644 1928
Practical bee-breeding./London, Putnam, 1928./248 p. illus./B

GLOUCESTERSHIRE BEE-KEEPERS ASSOCIATION 645 [1974]
The scrap book of the Gloucestershire Bee-keepers Association. The British Bee-keepers Association centenary celebration 1874–1974./[Hartpury (Glos.)], Gloucestershire Bee-keepers Association, [1974]./2 leaves, 32 p. 2 pl. 30 cm./B

GOAMAN, Muriel 646 1954
Judy's and Andrew's book of bees./London, Faber & Faber, 1954./39 p. illus. by Hazel Cook./B

For children./Hodgson 216.

GOUDEY, Alice E. 647 1962
Here come the bees!/London, Macmillan, 1962./94 p. illus. by Garry Mackenzie./B

For children./Hodgson 155.
Pub. in U.S.A. 1960./Johansson 191.

GRÉE, Alain 648 1973
Little Tom finds out about bees./London, Harrap, 1973./22 p. illus. by Gérard Grée. 25 × 21 cm./B

Little Tom Series.

For children./Hodgson 103.

Translated & adapted by Denise Lebreton from *Petit Tom et son amie l'abeille* 1971.

GREEN, W. H. Bassett 648a 1944
The queen of the golden bees./Bristol, Arrowsmith, 1944./71 p. illus. 23 pl. incl. front. by M. Forster Knight, incl. in pagination.

For children.

GREGG, A. L. 649 1949
Philosophy and practice of bee-keeping./Petts Wood (Kent), Bee Craft, 1949./240 p. 8 pl. front. bibliog./B

GROUT, Roy A. *editor* 650 1947
The hive and honey bee: a new book on beekeeping to succeed the book 'Langstroth on the hive and the honey bee'./London, Chapman Hall, 1947./633 p. illus. bibliog./A

Pub. in U.S.A. 1946./Johansson 198 gives details of later eds.

Edited by Roy A. Grout assisted by the staff of *American Bee Journal.*

HAMILTON, William 651 1945
The art of bee-keeping./York, Herald Printing Works, 1945./2 leaves, 182 p. 12 pl. front./B

2nd rev. & enlarged ed./York, Herald Printing Works, 1951./2 leaves, 200 p. 12 pl. front./B

HARKER, Leonard S 652 1938
Blazing the trail: reminiscences of A. Z. Abushâdy, poet—bee-master—humanist./London, C. W. Daniel, 1938./137 p. 8 leaves, 71 pl. bibliog./B

Dr. Abushâdy (1892–1955) promoted various beekeeping developments in England and Egypt.

HARWOOD, A. F. 653 [1947]
British bee plants./[Foxton (Herts.)], Apis Club, [1947]./81 p. 2 leaves./B

HAWES, Judy 654 1968
Watch honeybees with me./London, Black, 1968./[40] p. illus. by Helen Stone, 19 × 22 cm./B
Let's Read-and-find-out Science Books.
For children./Hodgson 121.
Pub. in U.S.A. 1964./Johansson 214.

HAWES, Judy 655 1969
Bees and beelines./London, Black, 1969./[40] p. illus. by Aliki. 20 × 22 cm.
Let's Read-and-find-out Science Books.
For children./Hodgson 116.
Pub. in U.S.A. 1964./Johansson 213.

HEARD, Gerald *pseud.*
(HEARD, Henry Fitzgerald) 656 1942
A taste for honey./London, Cassell, 1942./v, 153 p./A
Anr. ed./Harmondsworth (Middx), Penguin Books, 1961./A
Pub. in U.S.A. 1941.
Fiction.

HEATH, Ambrose 657 1956
Honey cookery./London, Neville Spearman, 1956./88 p./B

HERROD-HEMPSALL, William 1929
Bee-keeping simplified for the cottager and smallholder. (4th ed. of 471.)
See also 401, 458, 474, 480.

HERROD-HEMPSALL, William 658 1930, 1937
Bee-keeping new and old: described with pen and camera.
Vol. I. London, British Bee Journal, 1930./5 leaves, 772 p. illus. front./B
Vol. II. London, British Bee Journal, 1937./4 leaves, pp. 773–1842. illus. front./B

HERROD-HEMPSALL, William 659 1938
Anatomy, physiology and natural history of the honey bee./London, British Bee Journal, 1938./2 leaves, 172 p. illus./B
2nd ed./London, British Bee Journal, 1943./2 leaves, 172 p. illus./B

HERROD-HEMPSALL, W[illiam] 660 1938
The bee-keeper's guide to the management of bees in movable comb hives./London, British Bee Journal, 1938./ix, 169 p. 4 leaves, illus. front./B
2nd ed./London, British Bee Journal, 1938.

3rd ed./London, British Bee Journal, 1940./ix, 169 p. illus. front./B
4th ed./London, British Bee Journal, 1941./xi, 169 p./B
5th ed./London, British Bee Journal, 1942.
6th rev. ed./London, British Bee Journal, 1943./3 leaves, 169 p. illus. front./B
7th ed./London, British Bee Journal, 1944.
8th rev. ed./London, British Bee Journal, 1947./168 p. illus./A

HESS, Gertrud 661 [1955]
The bee./London, Stanford, [1955]./68 p. illus./B
Hallwag Pocket Library No. 4.
For children./Hodgson 206.
Translated by W. Gossling from *Die Biene* 1955.

HERZOG, Arthur 662 1974
The swarm./London, Heinemann, 1974./256 p./B
Anr. ed./London, Pan Books in association with Heinemann, 1976./221 p./B
Pub. in U.S.A. 1974.
Fiction; also made into a film.

HILLYARD, T. N. & MARKHAM, J. 663 1968
A survey of beekeeping in Ireland./Dublin, An Foras Talúntais, Horticulture & Forestry Division, 1968./2 leaves, 49 p. 1 leaf./B

HOCKEY, S. F. *editor* 663a 1975
The account-book of Beaulieu Abbey./London, Royal Historical Society, 1975./vi + 348 p./B
Camden Fourth Series Vol. 16
Includes much information on honey and wax production around 1270.
See ms 8.

HODGES, Dorothy 664 1952
The pollen loads of the honeybee: a guide to their identification by colour and form. Containing a chart of pollen load colours recorded in the south of England./London, Bee Research Association, 1952./51 p. 5 leaves, illus. 35 pl. (5 col.), 6 charts of hand-mounted colours, bibliog./B
Reprint of some pages, with one of the colour plates, entitled *Pollen grain drawings from "The pollen loads of the honeybee"*./London, Bee Research Association, 1964./31 pl. (1 col.), 9 other leaves/B
Facsimile reprint of 1952 ed. 1974. with 14 p. and 1 pl. added, the 6 colour charts directly printed./B

HODGSON, Natalie B. 665 1973
Children's books on bees and beekeeping./London, Bee Research Association, 1973./iii, 55 p. 30 cm./B

Bee Research Association Bibliography No. 14.
Companion to *Guide to children's literature on bees and beekeeping* (Bee Research Association Reprint M70, *Bee World* 54 (3): 121–138 (1973).

HOGWOOD, A. M. 666 1956
Handbook of beekeeping by the natural method./Altrincham (Ches.), Sherratt, 1956./76 p. illus./A

Anr. ed./Altrincham (Ches.), Sherratt, 1957./A

HOOKE, Robert 1961
Micrographia restaurata. Reprint of 1745 ed. (*see* 40)

HOOPER, M. M. 667 [1939]
Common sense beekeeping./London, Link House Publications, [1939]./96 p. illus. front./B

2nd ed./London, Link House Publications 1943./78 p. illus./A
3rd ed./London, Link House Publications [1946]./78 p. illus. front./B
4th ed./London, Link House Publications [1948?]./78 p. illus. front./B

HOOPER, Ted 668 1976
Guide to bees and honey./Poole (Dorset), Blandford Press, 1976./260 p. illus. 2 pl. bibliog./B

HOPKINS, Isaac & MULVANY, T. J. 1930
Practical bee-keeping. (Anr. ed. of 519.)

HORTICULTURAL EDUCATION ASSOCIATION, FRUIT COMMITTEE 669 [1961]
The pollination of fruit crops./Canterbury (Kent), Gibbs, [1961]./68 p. illus. bibliog./B

Reprinted from Scientific Horticulture Vol. XIV & XV.

HOWES, F. N. 670 1945
Plants and beekeeping: an account of those plants, wild and cultivated, of value to the hive bee, and for honey production in the British Isles./London, Faber & Faber, 1945./224 p. 32 pl. bibliog./B

Reprinted 1946./B
Reprinted 1948.

HOYT, Murray 671 1966

The world of bees./London, Bodley Head, 1966./254 p. 8 pl./B

Pub. in U.S.A. 1965./Johansson 236.

HUGHES, Jill 672 1976

A closer look at bees and wasps./London, Hamish Hamilton, 1976./30 p. illus. by Tony Swift & Norman Weaver. 26 × 19 cm./B

For children.

IMMS, A. D. 673 1938

Social behaviour in insects./2nd ed. London, Methuen, 1938./xi, 117 p. illus./B

Methuen's Monographs on Biological Subjects; 1st ed. not traced.

IMPERIAL ECONOMIC COMMITTEE 674 1928

Report of the Imperial Economic Committee on marketing and preparing for market of foodstuffs produced within the Empire. Sixth report—poultry and eggs, and seventh report—honey./London, H.M. Stationery Office, 1928./83 p./B

IRELAND.
DEPARTMENT OF AGRICULTURE 1931, 1941, 1943, 1946

Instruction in bee-keeping for the use of Irish bee-keepers. (Other eds of 432.)

IRELAND.
DEPARTMENT OF AGRICULTURE 675 [1941]

Hints on bee-keeping./Dublin, Department of Agriculture, [1941]./16 p. 1 pull-out pl./B

Anr. ed./Dublin, Department of Agriculture, 1957./23 p. 3 pull-out pl./B

Eire Department of Agriculture Leaflet No. 34.

IRELAND.
DEPARTMENT OF AGRICULTURE 676 1945

Horticulture and bee-keeping./Dublin, Stationery Office, 1945./234 p. illus. 30 pl. incl. pull-out./B

IRELAND.
DEPARTMENT OF AGRICULTURE 677 1958

Irish bee-keepers' manual./Dublin, Stationery Office, 1958./105 p. 67 pl. +2 pull-out pl./B

Successor to *Instruction in bee-keeping for the use of Irish bee-keepers* 1905 (*see* 432), with many of the same illustrations but the text rewritten. No later edition up to 1976; often out of print.

JAMES, E. L. B. 678 [1950]

Bee-keeping for beginners and others./[Birmingham, Author, 1950]./xvi, 281 p. XLI pl. incl. in pagination./B

Hodgson 246.

JONES, J. E., GRAHAM, A. J.
& SACKETT, L. H. 679 [1974]

An Attic country house below the Cave of Pan at Vari./London, Thames & Hudson & Athens, British School at Athens, [1974]./[98 p.] illus. 24 pl./B

Originally published in *The annual of the British School at Athens* Vol. 68 pp. 355–452, pl. 63–86, 1973.

Terracotta hives were found, the first known from Ancient Greece.

JONES, J. Evans 680 1960

Rhamant y gwenyn. [The romance of bees]./Llandybie (Carms.), Llyfrau'r Dryw, 1960./136 p. illus./A
In Welsh.

KAUFMAN, Mervyn [D.] 681 1968

Honeybees./London, Angus & Robertson, 1968./64 p. illus. by Douglas Howland./B

For children./Hodgson 136.
Pub. in U.S.A. as *No holidays for honeybees* 1966./Johansson 258.

KELSEY, W[ilfred] Eric 682 1945

The spell of the honey bee./London, Chapman & Hall, 1945./vii, 263 p. illus. XLIX pl. incl. front./B

Rev. ed./London, Chapman & Hall, 1947./xi, 274 p./B
Anr. ed./London, Chapman & Hall, 1953.

KENLY, Julie Classon 683 1935
Cities of wax./London, D. Appleton-Century, 1935./250 p. illus./A
For children./Hodgson 301.
Pub. in U.S.A. 1935./Johansson 264.

(*cancelled*) 684

LADYMAN, Phyllis 685 [1971]
Honeybee and garden ant./London, Methuen, [1971]./23 p. illus. 14 cm./**B**
Stand and Stare Series No. 27.
For children./Hodgson 104.

LAIDLAW, Harry H. & ECKERT, J. E. 686 1962
Queen rearing./2nd ed. London, Cambridge University Press, 1962./viii, 165 p. illus. front./B
1st ed. pub. in U.S.A. 1950./Johansson 281.

LAVINE, Sigmund 687 1962
Wonders of the hive./Exeter (Devon), Wheaton, 1962./92 p. illus./B
For children./Hodgson 187.
Pub. in U.S.A. 1958./Johansson 290.

LAWSON, J[ohn] W. 688 1931
Honeycraft in theory and practice./London, Chapman & Hall, 1931./xii, 228 p. illus. 17 pl. front. bibliog./B
2nd ed./London, Chapman & Hall, 1940./G
3rd rev. ed./London, Chapman & Hall, 1945./xii, 164 p. illus. VI pl. incl. front. bibliog./B

LAWSON, William 1927
A new orchard and garden. Reprint of 1626 3rd ed. (*see* 20)

LEEMING, John F. 689 1936
Claudius the bee./London, Harrap, 1936./159 p. illus.
Reprinted 1952./127 p./B
For children./Hodgson 296.

LEEMING, John F. 689a 1937
Thanks to Claudius./London, Harrap, 1937./158 p. 1 leaf, illus.
For children. A sequel to 689.

LEVANDOVSKA, Cecilia *translator* 690 1964
The world of the bee./London, Heinemann, 1964./46 p. 1 leaf, illus. by Mateusz Gawryś./B

For children./Hodgson 181.

Translated by Cecilia Levandovska from *Wielkie prace małej pszczoły* by Nasza Księgarnia 1959.

LEWELLEN, John [Bryan] 691 [1959]
The junior true book of honeybees./London, Muller, [1959]./47 p. illus. by P. Jackson./A

For children./Hodgson 223.

Pub. in U.S.A. 1953 as *The true book of honey bees.*/Johansson 292.

LIGHTNING, R. H. 692 1968
A biographical sketch of Charles Nash Abbott (1830–1894), the beemaster of Southall. A noted resident of Hanwell, Southall and Ealing./London, Ealing Local History Society, 1968./1 leaf, 69 p. illus./B

See 336, *Beekeeping for the million* by C. N. Abbott 1876; C. P. Abbott (author of 532) was his grandson.

LINDAUER, Martin 693 1961
Communication among social bees./London, Oxford University Press, 1961./x, 1 leaf, 143 p. illus. bibliog./B

Pub. in U.S.A. 1961./Harvard Books in Biology No. 2./Johansson 300.

LISNEY, Amy F. E. 694 1953
The bee walk: being the romance and practice of beekeeping./[Bray (Co. Wicklow)], Author, 1953./114 p. 17 pl. front./B

LONGFORD, Herbert G. 695 [1946]
The castle that hung in the air./Birmingham, Cornish Bros. [1946]./39 p. illus./A

For children./Hodgson 263.

LUND, Harold J. 696 1947
A man and his bees: a bee-keeper's chronicle./London, Sidgwick & Jackson, 1947./6 leaves, 156 p. 1 leaf./B

LYALL, Neil & CHAPMAN, Robert 697 1976
The secret of staying young: Pollen-B and how it can prolong your youth. London, Pan Books, 1976./160 p. illus. 8 pl. bibliog./B

Pollen-B is a proprietary pollen preparation.

MACE, Herbert [1927]
A book about the bee. (anr. ed. of 493, *see also* 505, 520.)

(*cancelled*) 698

MACE, Herbert 699 1927
Modern bee-keeping./Harlow (Essex), Author, 1927./225 p. illus. 16 pl./B

Rev. ed./Harlow (Essex), Bee-keeping Annual Office, 1933./225 p. 16 pl./B

MACE, Herbert 700 1928
Beekeepers' associations. A critical survey./Harlow (Essex), Author, 1928./34 p./B

MACE, Herbert 701 [1930]
Bee matters and beemasters: with an appendix 'Who's who in bee-keeping'. Harlow (Essex), Beekeeping Annual, [1930]./144 p. illus. 14 pl. front./B

Anr. ed./Harlow (Essex), Beekeeping Annual [1931]./152 p./B
Anr. ed./Harlow (Essex), Beekeeping Annual 1944./A

MACE, Herbert 1930
A book about the bee. (Anr. ed. of 493).

MACE, Herbert 702 1936
Bee farming in Britain./Harlow (Essex), Bee-keeping Annual, 1936./108 p. illus. VIII pl. bibliog./B

MACE, Herbert 703 [1942]
Beekeeping in simplicity: a guide for the beginner./Harlow (Essex), Beekeeping Annual, [1942]./16 p. bibliog./B

MACE, Herbert 704 [1944]
Bee gardens./Harlow (Essex), Beekeeping Annual, [1944]./16 p./B

MACE, Herbert 705 1949
Bees, flowers and fruit: the story of insect-plant relations./Harlow (Essex), Beekeeping Annual, 1949./184 p. illus. bibliog./B

MACE, Herbert 706 1952
The bee-keeper's handbook./London, Ward Lock, 1952./256 p. illus. 14 pl. front. bibliog./B

Rev. ed./London, Ward Lock, 1976./*The complete handbook of bee-keeping* prepared by Owen Meyer./192 p. illus. bibliog./B

McFADDEN, Audrey G[reenhough] *compiler* 707 [1963]
101 ways of using honey./Dublin, Chivers & Sons, [1963]./32 p. illus./B

MACFIE, D. T. 708 [1936]
Practical bee keeping and honey production./London, Collingridge, [1936]./[120] p. (numbered 124), 8 pl./B

Anr. ed./London, Collingridge, [n.d.]./127 p. 7 pl. front./B

MACKENSEN, Otto & ROBERTS, W. C. 709 [1949]
A manual for the artificial insemination of queen bees./Kings Langley (Herts.), Badgerdell Apiaries, [1949]./31 p. illus. bibliog. 25 cm./B

Pub. in U.S.A. 1948./Johansson F73.

McNICOL, Harry 710 1953
The young beekeeper./London & New York, Warne, 1953./127 p. illus./B

For children./Hodgson 225.

MACRAE, F. B. 711 1941
Bees and how to handle them./London & Redhill (Surrey), United Society for Christian Literature & Lutterworth Press, 1941./48 p. illus./B

'Written for general reading in . . . native territories in Africa.'

MAETERLINCK, Maurice 1927–1948
The life of the bee. (Other eds of 422)

MAETERLINCK, Maurice 1934–1949
The children's life of the bee. (Other eds of 490)

MALYSHEV, S. I. 712 1968
Genesis of the Hymenoptera and the phases of their evolution./London, Methuen, 1968./viii, 319 p. illus. front. bibliog./B

Translated by National Lending Library for Science and Technology from *Stanovlenie pereponchatokrylykh i fazy ikh zvolyutsii* 1966.

[MANDEVILLE, Bernard de] 1934 & 1970
The fable of the bees. (*see* 76)

MANLEY, R[obert] O[rlando] B[eater] 713 1936
Honey production in the British Isles./Reading & London, Bradley, 1936./343 p. XV pl. incl. front./B

Anr. ed./London, Faber & Faber, 1946./328 p. 14 pl. front./B

MANLEY, R[obert] O[rlando] B[eater] 714 1946
Honey farming. / London, Faber & Faber, 1946. / 293 p. 14 pl. front. / B
2nd impression, 1946, lacks photographs numbered 28 & 29 in 1st impression, which lacks photographs numbered 23 & 25 in 2nd; both have 29 in all
Anr. ed. 1951.

MANLEY, R[obert] O[rlando] B[eater] 715 1948
Bee-keeping in Britain. / London, Faber & Faber, 1948. / 439 p. 30 pl. (2 col.) col. front. bibliog. / B

MAPLET, John 1930
A greene forest. Reprint of 1614 ed. (*See* 6)

MAXWELL, Gordon Stanley 716 [1931]
Wondrous workers tell their story. A visit to a bee hive and an ants' nest. / London, Brodie, [1931]. / 47 p. illus. / A
Anr. ed. / London, Brodie, [c. 1950]. / A
Brodie Books No. 76. / In the beehive pp. 7–21.
For children. / Hodgson 309.

MAYNE, William [James Carter] 717 1955
A swarm in May. / Oxford, Oxford University Press, 1955. / viii, 199 p. illus. by C. Walter Hodges. / B
Anr. ed. / Oxford, Oxford Children's Library, 1962. / vi, 199 p.
Fiction, for children. / Hodgson 208.

METCALF, F. H. 718 1948
The bee community: the study of an insect. / Petts Wood (Kent), Bee Craft, 1948. / 102 p. illus. vi pl. (3 col.) / B
Hodgson 259.

MASCALL, Leonard 1934
The countryman's jewel. (*See* 8)

METHLEY, V. M. & others 719 1939
What the bees heard and other stories. / London & Glasgow, Collins, 1939 / 96 p. illus. / B
Collins Clear Type Supplementaries Grade 5 No. 6. / Title story pp. 5–18.
For children. / Hodgson 284.

MILNE, A. A. 720 1970

Pooh hears a buzzing noise and meets some bees./London, Methuen, 1970./47 p. illus. by Ernest H. Shepard. 9 × 6 cm./B

Part of Winnie-the-Pooh, by A. A. Milne 1926./illus. by Ernest H. Shepard. Boxed with three other stories as a set, under general title *Pooh's pot o'honey*.
For children.

MILNE, A. A. 721 1975

Winnie-the-Pooh and some bees./London, Methuen, 1975./[28] p. illus. by E[rnest] H. Shepard./B

This is the same as 720, in a different form.

For children./Hodgson 316.

MINISTRY OF AGRICULTURE
see **BOARD OF AGRICULTURE** 456, 494

MINISTRY OF AGRICULTURE
& FISHERIES 722 1927

Collected leaflets on bee keeping./3rd ed. London, H.M. Stationery Office, 1927./49 p. illus. 5 pl./B

Sectional Vol. No. 7.
For earlier eds see Board of Agriculture (494); for later eds see 723.

MINISTRY OF AGRICULTURE
& FISHERIES 723 1930

Bee-keeping./4th ed. London, H.M. Stationery Office, 1930/v, 55 p. illus. 7 pl./B

5th ed./London, H.M. Stationery Office, 1932./53 p./A
6th ed./London, H.M. Stationery Office, 1937./iii, 56 p. illus. 6 pl./B
Reprinted 1940./iii, 54 p. illus. 7 pl./B
Reprinted with amendments 1941./iii, 52 p. illus. 8 pl./B
Reprinted with further amendments 1943./ii, 44 p. illus. 8 pl./B
7th ed. by Colin G. Butler./London, H.M. Stationery Office, 1945./28 p. illus./B
Reprinted 1946./B
8th ed. by Colin G. Butler./London, H.M. Stationery Office, 1951./28 p. illus./B

9th ed. by Colin G. Butler./London, H.M. Stationery Office, 1957./28 p. illus./B
10th ed./London, H.M. Stationery Office, 1966./B
Reprinted 1968./v, 25 p./B
11th ed./London, H.M. Stationery Office,/v, 25 p./B
Ministry of Agriculture, Fisheries & Food Bulletin No. 9; for earlier eds see 722, 494.

MINISTRY OF AGRICULTURE & FISHERIES 724 1931
The marketing of honey and beeswax in England and Wales./London, H.M. Stationery Office, 1931./92 p. illus./B

MINISTRY OF AGRICULTURE & FISHERIES 725 [1932]
Honey: grading and marking./London, H.M. Stationery Office, [1932]./ 15 p. 2 pl./B
Anr. ed./London, H.M. Stationery Office, 1935./16 p. illus./B
Anr. ed./London, H.M. Stationery Office, 1938./16 p. illus./B
Marketing Leaflet No. 31.

MINISTRY OF AGRICULTURE & FISHERIES 726 1937
Cowan Memorial Library. Hand list of books in English printed after 1830./London, Ministry of Agriculture & Fisheries, 1937./35 p./B

MINISTRY OF AGRICULTURE & FISHERIES 727 1945
Diseases of bees./London, H.M. Stationery Office, 1945./25 p. 4 pl./B
Reprinted 1946, 1948 & 1949.
2nd ed./London, H.M. Stationery Office, 1951./28 p. 8 pl./A
3rd ed./London, H.M. Stationery Office, 1959./iv, 31 p. 12 pl./B
4th ed./London, H.M. Stationery Office, 1969./iv, 34 p. XII pl./B
Reprinted 1971./B
Ministry of Agriculture & Fisheries Bulletin No. 100.

MINISTRY OF AGRICULTURE & FISHERIES 728 1953
Honey from hive to market./London, H.M. Stationery Office, 1953./iv, 30 p. illus. 8 pl./B
2nd rev. ed./London, H.M. Stationery Office, 1963./vi, 28 p. illus. 8 pl./B
Ministry of Agriculture & Fisheries Bulletin No. 134.

MINISTRY OF AGRICULTURE, FISHERIES & FOOD 729 1964

Beehives./3rd ed. London, H.M. Stationery Office, 1964./iv, 23 p. illus. 4 pl./B

4th ed./London, H.M. Stationery Office, 1968./iv, 27 p. illus. 4 pl./B

Ministry of Agriculture, Fisheries & Food Bulletin No. 144.
See also Bee-hives 576, by Colin G. Butler 1949.

MINISTRY OF AGRICULTURE, FISHERIES & FOOD 730 1970

Swarming of bees./London, H.M. Stationery Office, 1970./iv, 38 p. bibliog./B

Ministry of Agriculture, Fisheries & Food Bulletin No. 206.

MORE, Daphne 731 1976

The bee book: the history and natural history of the honeybee./Newton Abbot (Devon), David & Charles, 1976./143 p. illus. bibliog./B

MORSE, Roger A. 732 1973

The complete guide to beekeeping./London, Pelham Books, 1973./204 p. 2 leaves, illus./B

Pub. in U.S.A. 1972.

MORSE, Roger A. 733 1975

Bees and beekeeping./London, Cornell University Press, 1975./295 p. 4 leaves. illus. bibliog./B

MOUNTAIN, M. F. 734 1965

Trees and shrubs valuable to bees./London, Bee Research Association, 1965./15 p bibliog./B

Anr. ed. with minor revisions./London, Bee Research Association, 1975./14 p./B

NIXON, Gilbert [E. J.] 735 1954

The world of bees./London, Hutchinson, 1954./214 p. illus. by A. Smith, 16 pl. front./B

Hutchinson's Nature Library.

Anr. ed./London, Arrow Books, 1959./192 p. 16 pl./B
Grey Arrow Books No. 19.

For children./Hodgson 219.

NORRIS, P. E. 736 1954
About honey: nature's elixir for health and energy. / London, Thorsons, 1954. / 72 p.
Reprinted 1955, 1956.

OLDFIELD, J[osiah] [1952]
Honey for health. (Anr. ed. of 491)

OWEN, Julia 737 1965
Doctors without shame. / Bromley (Kent), Julia Owen Publications, 1965. / 254 p. 1 leaf. / (B)
Advertised before publication apparently with the title *A sting in the tale.* Anti-establishment approach to bee venom treatment for rheumatoid arthritis, as also the author's *Treat yourself for your rheumatic or arthritic disease and avoid the doctor* [1962] and *Clamouring at the citadel* 1963.

OXFORD SCIENTIFIC FILMS *editor* 738 1976
Bees and honey. / London, Whizzard Publications in association with André Deutsch, 1976. / [28] p. of col. photographs by David Thompson, 19 × 24 cm. / B
Nature's Way Series No. 3.
For children.

PALMER, R. 739 1940
Bombus, the bumble-bee. / London, Nelson, 1940. / A
Animals All.
For children. / Hodgson 281.

PARNWELL, Eric Cecil 740 1937
The life of the honey bee. / London, Oxford University Press, 1937. / 52 p. / A
Simple Science in Simple English.
For children. / Hodgson 293.
Adapted from *The lore of the honey-bee* by Tickner Edwardes (440)

PAVORD, A. V. 741 1970
Bees and beekeeping. / London, Cassell, 1970. / 3 leaves, 123 p. illus. front. bibliog. / B
Cassell's Pet and Fancy Books No. 12.
Hodgson 108.

PELLETT, Frank C. [1928]
Productive bee-keeping. (4th rev. ed. of 476)

PEARSON, H. M. 742 [1955]
Bee stings and how to avoid them./London, British Bee Publications, [1955]./52 p. illus. bibliog./B

PHILLIPS, Everett Franklin 1937 & 1949
Beekeeping. (other eds. of 472)

PHYTHIAN, Dorothy 743 1933
The bee-garden./London, Herbert Jenkins, 1933./311 p. 4 leaves./B
Fiction.

PIM, Sheila 744 1952
A hive of suspects: an Irish detective story./London, Hodder & Stoughton, 1952./255 p./B
Fiction.

PLOT, Robert 1972
The natural history of Oxfordshire. Facsimile reprint of 1705 ed. (49)

PRICE, Joseph 745 [1949]
Price on beekeeping./Stafford, Staffordshire Beekeepers' Association, [1949]./70 p./B
Pub. posthumously, compiled by G. St. John Jones.

PROCTOR, Michael & YEO, Peter 746 1973
The pollination of flowers./London, Collins, 1973./418 p. illus. 56 pl. IV col. pl. bibliog./B
New Naturalist Series No. 54.

RANSOME, Hilda M. 747 1937
The sacred bee in ancient times and folklore./London, Allen & Unwin, 1937./308 p. illus. XII pl. incl. front. bibliog./B
Hodgson 294.
Pub. in U.S.A. 1937./Johansson 446.

RATCLIFF, John [H.] R. 748 1949
Beekeeper's folly./London, Macdonald, 1949./xvi, 270 p. 1 leaf, illus. 16 pl. front./B

RENDL, Georg 749 1933
The way of a bee./London, Longmans Green, 1933./vi, 1 leaf, 168 p. front./B
Hodgson 306.
Translated by Patrick Kirwan from *Der Bienenroman* 1931.

RENNIE, John 750 1927
Acarine disease in hive bees: its cause, nature, and control./Aberdeen, Milne & Hutchinson, 1927./34 p. illus. V pl. incl. front./B
Bulletin of the North of Scotland College of Agriculture No. 33.
John Rennie was also author of 499, 506.

REYNOLDS, John 751 1974
Bees and wasps./London, Priory Press, 1974./80 p. illus. bibliog./B
For children.

RIBBANDS, C. R[onald] 752 1953
The behaviour and social life of honeybees./London, Bee Research Association, 1953./352 p. illus. 8 pl. front. bibliog./B
Hodgson 226.

RICHARDS, O. W. 753 1953
The social insects./London, Macdonald, 1953./xiii, 15–219 p. Bees, pp. 76–118. illus. 14 pl. front. bibliog./B

RICHARDS, O. W. 754 1956
Hymenoptera: introduction and keys to families./London, Royal Entomological Society, 1956./94 p. illus. bibliog./B
Royal Entomological Society of London, Handbooks for the identification of British insects, Vol. VI, Part 1.

RICHARDSON, M. K. 755 [1968]
Benwig the bee./[London], St. Paul Publications, [1968]./32 p. col. illus./B
Benwig Books.
For children./Hodgson 109.

RICHES, H. R. C. 756 1976
Bee keeping./London, Foyles, 1976./92 p. illus. 4 col. pl. bibliog./B

RIMMER, Trevor 757 [1973]
History of Manx beekeeping./[Douglas?], Isle of Man Natural History and Antiquarian Society, [1973]./24 p. illus./B
Reprinted from the *Proceedings* of the Society.

RITCHIE, T. 758 [1942]
The bees./London, Chatto & Windus, [1942]./12 p. col. illus. 8 × 7 cm./B
Midget Books, Nature Stories No. G5.
For children./Hodgson 275.

ROELS, Iliane 759 1968
The bee./Edinburgh & London, W. & R. Chambers, 1968./20 p. illus. 18 × 20 cm./B
Animals at Home Series; translated by Marion Koenig.
For children./Hodgson 129 under Koenig, M.
Pub. in U.S.A. 1967./Johansson 275 under Koenig, M.

ROELS, Iliane 760 1976
The bee and the cherry-tree./Edinburgh, W. & R. Chambers, 1976./2 leaves, [21] p. illus. by Pauline Baynes, 18 × 20 cm./B
How Life Goes On Series; adapted from the text by Claude Nicolas.
For children.

ROHDE, E. S. 760a 1935
Shakespeare's wild flowers: fairy lore, gardens, herbs, gatherers of simples and bee lore./London, Medici Society, 1935./xi, 236 p. col. front. + 4 col. pl. and line drawings.
Bee lore pp. 197–215./(B)

ROLLE, Richard (*see ms* 10) 1930
Selected works. Anr. ed.

ROLLE, Richard (*see ms* 10) 1931
English writings.

ROOD, Ronald N. 761 [1966]
The how and why wonder book of ants and bees./London, Transworld, [1966]. 48 p. illus. by Cynthia Iliff Koehler and Alvin Koehler./B
Reprinted 1970, 1971, 1975.
For children.
Pub. in U.S.A. 1962./Hodgson 156./Johansson 460.

ROWSELL, Henry & MACFARLANE, Helen 762 1974
Henry's bee herbal: modern applications of honey therapy./Wellingborough (Northants.), Thorsons, 1974./128 p./B

RUSSELL, Sir E[dward] J[ohn] & others 762a 1935
The cause and control of swarming in bees: being the report of a conference held at Rothamsted . . ./Harpenden (Herts.), Rothamsted Experimental Station, 1935./31 p. illus./B

Rothamsted Conferences XX.
Contributions by Brother Adam, Miss A. D. Betts, C. F. Clay, W. Hamilton, D. Morland, W. H. J. Prior and L. E. Snelgrove.

RUSSELL, Sir E[dward] J[ohn] & others 762b 1936
Diseases of bees: being the report of a conference held at Rothamsted . . ./Harpenden (Herts.), Rothamsted Experimental Station, 1936./43 p. illus. 1 pl./B

Rothamsted Conferences XXII.
Contributions by Sir E. J. Russell, H. L. A. Tarr, G. D. Morison, O. Morgenthaler and J. I. Hambleton.

RUTLEY, Cecily M. 763 [1950]
Hum Hum a honey-bee./London, Warne, [1950]./15 p. illus./B

For children./Hodgson 250.

SCHOFIELD, A. Norman 764 1943
Teach yourself bee-keeping./London, Hodder & Stoughton, 1943./xii, 13–148 p. illus. by W. W. Newman, IV pl. bibliog./B

Teach Yourself Books.

2nd ed./London, English Universities Press, 1947./xii, 13–151 p. illus. IV pl. bibliog./B
Reprinted 1951.
Rev. ed./1958.
Anr. ed./London, Teach Yourself Books, 1971./xii, 13–151 p. illus. 4 pl. bibliog./B
Hodgson 274.
Pub. in U.S.A. 1947./Johansson 521.
Replaced by 806, by F. G. Vernon.

SCOTTISH BEEKEEPERS' ASSOCIATION 1930 & 1933
List of books in the library of the Association. (Supplements to 473.)

SCOTTISH BEEKEEPERS' ASSOCIATION 765 [1949]
Honey recipes./[Dundee], Scottish Beekeepers' Association, [1949]./16 p./B

SCOTTISH BEEKEEPERS' ASSOCIATION 766 [1949]
The Scottish beekeepers' handbook./[Dundee], Scottish Beekeepers' Association, [1949]./40 p. illus./B

SCOTTISH BEEKEEPERS' ASSOCIATION 1950
Catalogue of the Moir Library. (Anr. ed. of 473)

SCOTTISH BEEKEEPERS' ASSOCIATION 1965
Supplement 1950–1962. (to 1950 ed. of 473)

SCOTTISH BEEKEEPERS' ASSOCIATION 767 1973
An introduction to bees and beekeeping. [Ayr], Scottish Beekeepers' Association, [1973]./[36 p.] illus. bibliog./B

SECHRIST, Edward Lloyd 768 1958
Amateur beekeeping./London, Robert Hale, 1958./156 p. illus. bibliog./B
2nd ed./London, Robert Hale, 1976./156 p. 2 leaves, illus. bibliog./B
Pub. in U.S.A. 1955./Johansson 524.
U.K. ed. edited by A. C. Brown.

SHEEHAN, Angela 769 1976
The bumblebee./London, Angus & Robertson, 1976./[26] p. col. illus. by Maurice Pledger./B
Eye-view Library.
For children.

SHIDA, John 770 1976
Beekeeping./Gloucester, Thornhill Press, 1976./30 p. illus./B
A Thornhill Guide No. 4.

SIMMINS, Samuel 1928
Facts for beekeepers. (rev. ed. of 369)
A modern bee-farm . . . (rev. ed. of 395)

SIMPSON, Marie *translator* 771 1954
Some recent Russian researches on bees and beekeeping./London, Bee Research Association, 1954./36 p. illus./B
Articles originally pub. in *Pchelovodstvo.*

SINCLAIR, W. 772 1969
Life of the honey-bee./Loughborough (Leics.), Ladybird Books formerly Wills & Hepworth, 1969./52 p. col. illus. by Jill Payne./B
Ladybird Natural History Book.
For children./Hodgson 118.

SMILEY, Virginia Kester 773 1957
The buzzing bees./New York & London, Abelard-Schuman, 1957./1 leaf, [72] p. illus. by Bob Hodgell. 21 × 20 cm./B
For children./Hodgson 196.

SMITH, D[avid] A. *editor* 774 1966
John Evelyn's manuscript on bees from Elysium Britannicum./London, Bee Research Association, 1966./40 p. illus. bibliog./B
Reprinted from *Bee World* 46(4); 116–131(1965). Includes complete transcript of previously unpublished original, written about 1655. The complete unpublished *Elysium Britannicum* is 34.

SMITH, Francis G. 775 1960
Beekeeping in the tropics./London, Longmans Green, 1960./xvii, 265 p. 1 leaf, illus. 32 pl. 4 col. pl. bibliog./B
Tropical Agriculture Series.

SMITH, Francis G. 776 1963
Beekeeping, a beginner's guide to profitable honey and beeswax production./London, Oxford University Press, 1963./ix, 129 p. illus. front. incl. in pagination./B
Oxford Tropical Handbooks.

SMITH, Victor J. 777 1946
Down among the beefolk: pen-pictures for young people from a beekeeper's log book./London, Kingsgate Press, 1946./74 p. illus./B
For children./Hodgson 264.

SMITH, Victor J. 778 1950
Down where the bee-folk fly: pen pictures from an East Anglian landscape. London, Independent Press, 1950./188 p. illus. by T. K. Williams./B
For children.

SNELGROVE, Louis Edward 779 [1934]

Swarming: its control and prevention./[Bleadon (Somerset)], privately printed, [1934]./xi, 13–95 p. illus. front./B

2nd ed./[Bleadon (Somerset)], privately printed, 1935./xi, 13–96 p. illus. incl. front./B
3rd ed./[Bleadon (Somerset)], privately printed, 1936./viii, 9–96 p. illus. incl. front./B
4th ed.
5th ed./[Bleadon (Somerset)], 1941./viii, 9–98 p. 1 leaf, illus. incl. front./B
6th ed./Bleadon (Somerset), Miss I. Snelgrove, 1942./viii, 9–98 p. illus. incl. front.
7th ed./Bleadon (Somerset), Miss I. Snelgrove, 1943./viii, 9–100 p. illus. incl. front./B
8th ed./Bleadon (Somerset), Miss I. Snelgrove, 1945./viii, 9–100 p. illus. incl. front./B
9th ed./Bleadon (Somerset), Miss I. Snelgrove, 1946./viii, 9–100 p. illus. incl. front./B
10th ed./Bleadon (Somerset), Miss I. Snelgrove, 1955.
11th ed./Bleadon (Somerset), Miss I. Snelgrove, 1963./viii, 9–110 p. illus. incl. front./B

SNELGROVE, L. E. 780 1940

The introduction of queen bees./[Bleadon (Somerset), Author], 1940./xv, 17–205 p. XI pl. incl. front. bibliog./B

2nd ed./Bleadon (Somerset), Miss I. Snelgrove, 1943./xv, 17–205 p. XI pl. incl. front. bibliog./B
3rd ed./Bleadon (Somerset), Miss I. Snelgrove, 1948./xv, 17–205 p. XI pl. incl. front. bibliog./B

L. E. Snelgrove was also author of 460.

SNELGROVE, L. E. 781 1946

Queen rearing./Bleadon (Somerset), Miss I. Snelgrove, 1946./xiv, 15–344 p. XXXV pl. incl. in pagination. bibliog./B

2nd ed./Bleadon (Somerset), Miss I. Snelgrove, 1949.
3rd ed./Bleadon (Somerset), Miss I. Snelgrove, 1966./B

SNODGRASS, R. E. 782 1956

Anatomy of the honey bee./Ithaca & London, Comstock Publishing Associates, 1956./xiv, 1 leaf, 334 p. illus. bibliog./B

Reprinted 1976.

SNOW, Dorothea 783 [1965?]
Honey bear. / Brighton (Sussex), Litor, [1965?]. / [16] p. illus. 25 × 16 cm. / B
Turn-a-dial Story Book.
For children.
Contains only a passing reference to honey.

SOUTHERNE, Edmund 1934
A treatise concerning the right use and ordering of bees. Mimeograph of 1593 ed. (*see* 12)

SPILLER, John 784 [1952]
The advantages of the house apiary. / [Taunton (Somerset), Author, 1952]. / 2 leaves, 56 p. illus. 10 pl. front. / B

STADTLAENDER, Chris 785 1967
Honey cookery. / London, Thorsons, 1967. / 120 p. / B
Translated by Janet A. Young from *Koche mit Honig koche gesund* by Chris Stadtlaender 1965.

STEBBING, Lionel *editor* 786 1963
Honey as healer. / London, Emerson Press, 1963. / A
2nd ed., 1965, by Lionel Stebbing.
3rd rev. ed. London, Emerson Press, 1968 / 39 p. bibliog. / B
First published pseudonymously by A. E. Abbot *editor and compiler* (*see* 531).

STEINER, Rudolf 787 1933
Nine lectures on bees given in 1923 to the workmen at the Goetheanum. / London, Anthroposophical Agricultural Foundation, 1933. / 95 p. 5 pl. / B
Translated by Marna Pease & Carl Alexander Mirbt.

STEP, Edward 788 1932
Bees, wasps and ants, and allied insects of the British Isles. / London, Warne, 1932. / xxv, 238 p. illus. 111 pl. (some col.) incl. front. / B
Wayside and Woodland Series.

STOCKTON, Frank R. 789 1967
The bee-man of Orn. / London, Collins, 1967. / 46 p. 1 leaf, col. illus. by Maurice Sendak. / B
For children. / Hodgson 144.
First pub. in U.S.A. 1964.

STODART, Eleanor 790 1973
The bee./London, Angus & Robertson, 1973./32 p. photographs by Ederic Slater. 17 × 21 cm./B
For children.
Pub. in Australia 1973.

STRATTON-PORTER, Gene 1929 & 1969
The keeper of the bees./(Other eds. of 522).

STUART, F[rank] S. 791 1945
Bee-keeping practice./London, Pearson, 1945./212 p. illus. VIII pl./B
2nd ed./London, Pearson, 1946./212 p. illus. VIII pl./B
3rd ed./London, Pearson, 1947./216 p. illus. VIII pl./B

STUART, Frank S. 792 1947
City of the bees./London, Allen & Unwin, 1947./183 p./B
Fiction.

STURGES, Arthur M. 793 1927
Swarm control and comb honey production./Eastbourne (Sussex), East Dean Apiaries, 1927./61 p. 1 leaf, illus. 15 cm./B

SZABO, Tibor I. 794 1972
Dictionary of beekeeping terms with allied scientific terms. Giving translations from and into English–Hungarian./London, Bee Research Association, 1972./45 p. 30 cm./B
Supplement I to Vol. I-IV./See 596, 599, 600, 601.

TAYLOR, Eileen Saville 795 1958
The honey bee./London, Philip & Tacey, 1958./16 p. col. illus. by Peter Green./B
The Look Books. Nature Activity Readers.
For children./Hodgson 192.

TAYLOR, Richard 796 1976
The joys of beekeeping./London, Barrie & Jenkins, 1976./5 leaves, 166 p. illus. by Meri Shardin./B
Pub. in U.S.A. 1974.

TEALE, Edwin Way 797 1942
The golden throng: a book about bees. / London, Hale, 1942. / 160 p. 63 pl. by author, incl. front. bibliog. 25 cm. / B

Anr. ed. / London, Museum Press, 1946. / S
Anr. rev. ed. / 1961.
2nd ed. / London, Museum Press, 1964. / 216 p. 60 pl. & front. incl. in pagination. / B

Hodgson 282.
Pub. in U.S.A. 1940. / Johansson 557.

TENNANT, Kylie 798 1956
The honey flow. / London, Macmillan, 1956. / 2 leaves, 348 p. / B

Fiction, about Australian beekeeping.

TIBBETS, Albert B. 799 1963
The first book of bees. / London, E. Ward, 1963. / 69 p. illus. by Helene Carter. / B

For children. / Hodgson 235.
Pub. in U.S.A. 1952. / Johansson 566.

TILLY, Bertha *editor* 800 1973
Varro the farmer: a selection from the Res rusticae. / London, University Tutorial Press, 1973. / viii, 362 p. 1 leaf. 7 pl. front. bibliog. / B

Pp. 120–127, 294–311 deal with bees and beekeeping.

TINSLEY, Joseph 801 1945
Beekeeping. / Glasgow, Aird & Coghill, 1945. / 94 p. 1 leaf, illus. / B

2nd ed. / Welwyn (Herts.), Taylor, 1945. / 116 p. 2 leaves, illus. / B
Cover-title of IBRA Library copy is *Beekeeping up-to-date.*

J. Tinsley was also author of 445.

TONSLEY, Cecil 802 1969
Honey for health. / London, Universal-Tandem, 1969. / 127 p. 8 pl. / B

Rev. enl. ed. / London, Universal-Tandem, 1973. / 4 leaves, 9–124 p. 2 leaves, 8 pl. / B

TOPSELL, Edward 1967
Historie of four-footid beastes . . . Reprint of 1658 ed. (*see* 17).

TUCK, M. 803 [1930?]
Tried and tested recipes for the use of honey and beeswax as used by beekeepers in Devonshire./Newton Abbot (Devon), Devon Press, [1930?]./19 p. 13 cm.

Reprinted Uxbridge, [for Robert] Lee, [n.d.]/19 p.

TUSSER, Thomas 1931 & 1954
Five hundred pointes of good husbandrie. (*See* 3)

USBORNE, Peter *editor* 804 1971
Bees./London, Macdonald, 1971./25 p. col. illus. by Tancy Baran./B
Macdonald Starters No. 26.
For children./Hodgson 106.

VANN, E. M. & VANN, C R. 805 1952
Honey bees./London, Taylor's Foreign Press, 1952./16 p. illus./A
G.B. Instructional Chart Books.
Anr. ed./London, Pitman, [1953]./16 p. 1 pl. (pull-out)./B
Childrens Nature Series No. 2.
For children./Hodgson 231.

VERNON, Frank [G.] 806 1976
Beekeeping./London, Hodder & Stoughton, 1976./xii, 212 p. illus. VIII pl. bibliog./B
Teach Yourself Books, successor to 764, by A. Norman Schofield.

VESEY-FITZGERALD, Brian 807 1969
The worlds of ants, bees and wasps./London, Pelham Books, 1969./117 p. illus. 16 pl./B

VIRGIL (Publius Vergilius Maro) 808 1940
The Georgics of Virgil./London, Jonathan Cape, 1940./95 p./B
Reprinted 1941, 1943, 1946.
Translated by C[ecil] Day Lewis.

VIVIAN, Francis *pseud.* **(ASHLEY, Arthur Ernest)** 809 1950
The singing masons./London, Hodder & Stoughton, 1950./256 p./A

WADEY, H. J. 810 1943
The bee craftsman: a short guide to the life story and management of the honey-bee./Petts Wood (Kent), Bee Craft, 1943./115 p. 8 pl. bibliog./B
2nd ed./Petts Wood (Kent), Bee Craft, 1944./116 p. illus./B
3rd ed./Petts Wood (Kent), Bee Craft, 1945./116 p. illus./B
4th ed./Petts Wood (Kent), Bee Craft, 1947./117 p. illus./B
5th ed./Petts Wood (Kent), Bee Craft, 1949./117 p. illus./B

WADEY, H. J. 811 [1946]
Introducing bee-keeping: a book for the novice./Bognor Regis (Sussex), Crowther, [1946]./74 p. 8 pl. front. 15 cm./B

Hodgson 266.

WADEY, H. J. 812 1948
The behaviour of bees and of bee-keepers./[Petts Wood (Kent)], Bee Craft, 1948./vii, 1 leaf, 11–175 p. illus. by Neil Nettleton./B

WAINE, Allan C. 813 1955
Background to beekeeping./London, Collingridge, 1955./120 p. illus. by David Hutter, 16 pl. bibliog./B

2nd rev. ed./Warminster (Wilts.), Bee Books New & Old, 1975./118 p./B

Hodgson 211.

WALKER, H. J. O. 814 1929
Descriptive catalogue of bee-books collected and offered for sale by Lt. Col. H. J. O. Walker./[Budleigh Salterton (Devon), Author], 1929./144 p./B

H. J. O. Walker is also author of 362, 445.

WALL, Dorothy 815 1934
The tale of Bridget and the bees./London, Methuen, 1934./55 p. illus./A

For children./Hodgson 302.

[WARWICKSHIRE BEEKEEPERS' ASSOCIATION] 816 [1976]
Warwickshire honey and wax recipes./Warwickshire, Warwickshire Beekeepers' Association, [1976]./20 p./B

WEBB, Addison 817 1945
Beekeeping for profit and pleasure./London, Museum Press, 1945./160 p. illus. by Natalie H. Davis & F. Martin Duncan./B

Anr. ed./1947./S

Edited & with a foreword by F. Martin Duncan.

WEBSTER, W. B. 1929, 1938, [1947]
The book of bee-keeping. (6th, 7th, & 8th eds of 401)

WEDMORE, E. B. 818 1932
A manual of bee-keeping for English-speaking bee-keepers./London, Arnold, 1932./xxiv, 413 p. illus. 7 pl. front./B

Reprinted 1942 & 1943./8 pl./B

2nd rev. ed./London, Arnold, 1945./xxiv, 389 p. illus. IX pl. incl. front./B

Facsimile reprint of 2nd ed. without pl./Warminster (Wilts.), Bee Books New and Old, 1975./B

WEDMORE, E. B. 819 1940
Electric heating of beehives. Preliminary report./London, British Electrical and Allied Industries Research Association, 1940./19 p. 5 pl. 27 cm./B

Technical Report W/T 5 of the Association.

Second report./1946./W/T 13./B

WEDMORE, E. B. 820 [1946]
Successful bee-keeping./London, Foyle, [1946]./163 p. illus. IV pl./B

WEDMORE, E. B. 821 1947
The ventilation of bee-hives./Petts Wood (Kent), Bee Craft, 1947./116 p. illus. 2 pl./B

WEIGHTMAN, Colin 822 1961
The border bees: anecdotes and memoirs of twenty years, 1940–1960./Consett (Co. Durham), Ramsden William Publications, 1961./[xi], 96 p. 1 leaf, illus. 8 pl. front./B

WHEELER, A. G. 823 [1955?]
An approach to beekeeping./Leyland (Lancs.), Ditchfield's British Books, [1955?]./32 p. 13 cm./B

Ditchfield's Little Wonder Book No. 31.

WHISTON, John W. 824 1976
History of the Staffordshire Beekeepers' Associations 1876–1976./Walsall (Staffs.), Staffordshire Beekeepers' Associations, 1976./85 p. 7 pl./B

WHITCOMBE, Harry J.
with DOUGLAS, John Scott 825 1956
Bees are my business./London, Gollancz, 1956./vii, 245 p. 8 pl./B
Pub. in U.S.A. 1955./Johansson 595.

WHITE, T. H. 826 1954
The book of beasts: being a translation from a Latin bestiary of the twelfth century./London, Jonathan Cape, 1954./296 p. illus./Bees pp. 153–159./A(B)
See ms 4.

WHITEHEAD, Stanley B. 827 1946
Honey bees and their management./London, Faber & Faber, 1946./153 p. 48 pl. bibliog./B

WHITEHEAD, Stanley B. 828 1954
Bees to the heather./London, Faber & Faber, 1954./96 p. illus. 8 pl./B

WILDMAN, Thomas 1970
A treatise on the management of bees. Reprint of 1770 2nd ed. (*see* 119)

WILLIAMS, Canning 829 1928
The story of the hive: a bee-lover's book./London, Black, 1928./viii, 200 p. front./B
Hodgson 311.

WILLIAMS, Melfyn R. 830 1972
Y fêl ynys. [The honey isle]./Pontypridd (Glam.) & Liverpool, Cyhoeddiadau Modern Cymreig, 1972./3 leaves, 135 p. illus. front./B
In Welsh.

WODEHOUSE, R. P. 831 1935
Pollen grains: their structure, identification and significance in science and medicine./London & New York, McGraw Hill, 1935./xv, 574 p. illus./B

WORTHY, W. 832 [1964]
All about a jar of honey./[London, Longmans Green, 1964]./32 p. col. illus. by Donald Dosset./B
All About Books No. 6.
For children./Hodgson 145.

Metaphorical titles

1662 to 1974

We ourselves have been misled upon occasion by titles of books in which the word bee or queen, honey or hive, is used metaphorically. In such books the bee often represents industry, and the hive signifies a choice collection of writings; honey seems to be used to imply sweetness, whether or not the contents bear this out.

The list printed here may perhaps save others spending time following a false trail; perhaps it will also inspire someone to search further for such titles.

[HORSEMAN, Nicholas]? *mt* 1 1662

The spiritual bee, or A miscellaney of scriptural, historical, natural observations, and occasional occurrencyes applyed in Divine meditations. By an University Pen./Oxford, A. & L. Lichfield for Edw. & Joh. Forrest, 1662./6 leaves, 135 p. 4 leaves, 12 cm./A
Anr. ed./Oxford, Crosley, 1667./257 p./A
Anr. ed./1822.
Extracts from the spiritual bee./London, Phillips, 1823./xiv, 152 p./A

mt 2 [c.1700]

The bee, a new song sung in the public gardens./[c.1700]/1 leaf, 33 × 11 cm. 3 verses of 8 lines./A
The words of the song are reproduced in Fig. 17; there were illustrations, not of bees, above and below the verses.

The Bee, or Universal Weekly Pamphlet, published a somewhat similar ditty, *The bee lover—a song*, about 1732 (Vol. 3, p. 1326). Here, although the writer 'from flower to flower still wanders free', he says in conclusion

These are but visits which I pay,
My Chloe's still my home.

mt 3 1715

The bee, a collection of choice poems./Pt. 1, 2, London, printed by T. Ilive; sold by S. Popping and A. Boulter, 1715./A

mt 4 [1790?]

The busy bee, or, Vocal repository. Being a selection of the most favourite songs, &c. contained in the English operas, that have been sung at the

THE

B E E.

A new Song

Sung in the Public Gardens

A Buſy humble Bee am I,
That range the garden ſunny,
From flower to flower I changing fly,
And every flower's my honey.
Bright Chloe with her golden hair,
A while my rich Jonquil is,
'Till cloy'd with ſipping Nector there,
I ſhift to roſy Phillis.
I ſhift, &c.

But Phillis's ſweet opening breaſt,
Remains not long my ſtation;
For Kitty now muſt be addreſt,
My pretty ſpic'd Carnation:
Yet Kitty's fragrant bed I'd leave,
To other flowers I'm rover;
And all in turns my love receive,
The gay wide garden over.
The gay, &c,]

Variety that knows no bounds,
My roving fancy edges,
And oft with Flora I am found,
In dalliance under hedges.
For as I am an errant B E E,
Who range each bank that's ſunny,
Both fields and gardens are my fee,
And every flower my honey.
And every, &c.

public gardens and written for select societies; together with an extensive collection of hunting songs, and a variety of Scotch and Irish ballads, &c./London, J. S. Barr, [1790?]/3 vol. xvi, 350 p., xii, 348 p., viii, 348 p./A

ANDERSON, James, LLD *mt* 5 1791–93

The bee or Literary weekly intelligencer, consisting of original pieces, and selections from performances of merit foreign and domestic./Edinburgh, Mundell, 1791–93/3 vol./A

mt 6 1795

The bee, a selection of poetry from the best authors. A new ed./London, Darton & Harvey, 1795./5 leaves, 191 p. engraving of skep on title page, 12 cm./A

New ed./London, Darton & Harvey, 1797./204 p./A G

6th ed./1811./vii, 204 p./A

Victoria and Albert Museum has an ed. dated 1822.

mt 7 1795

The Caledonian bee: or A select collection of interesting extracts, from modern publications with elegant copperplates./Perth, Morison, also London, Vernor & Hood, 1795./iv, 280 p. illus. 16 cm./A

mt 8 1796

The hive, a selection from modern writers; in prose and verse, embellished with elegant engravings./Edinburgh, Morison also Perth, Carlisle, York, London, 1796./3 leaves, 224 p. illus. 14 cm./A

mt 9 [1800?]

The busy bee, being a selection of most of the monstrous good songs, now singing at the different places of polite and public amusement./London, J. Evans, [1800?]./8 p./A

Fig. 17. This broadsheet was discovered in the British Library, which gives it the date 1700(?). We have failed to find any information about its composer or tune, or about its popularity, but see *mt* 2.

MILLINGEN, J. C. *mt* 10 1811

The bee-hive, a musical farce, in two acts, as performed at the Theatre Royal Lyceum./Music by Mr. Horn./London, 1811./

This entry is derived from Halkett and Laing's *Dictionary of anonymous and pseudonymous English literature* new and enl. ed. 1926. We have not seen the book.

HODGSON, Solomon *editor* *mt* 11 1812

The hive of ancient and modern literature: a collection of essays, narratives, allegories and instructive composition selected by the late Sol Hodgson. 4th ed./Newcastle, Hodgson, 1812./vii, 338 p. illus. by Bewick./ (A has 3rd ed. Newcastle 1806)

For use in schools. Not about bees, except that p. 87 has the story of *The two bees*.

mt 12 1824

The bee; a collection of poems chiefly designed for the young./Dublin, Jones, 1824./viii, 180 p. front. of the family at home, skep on title page./A

Anr. ed./Dublin, Society for Promoting the United and Scriptural Education of the Poor of Ireland, [1825?]./A

THOMSON, ? *mt* 13 1831

The bees and the drones. [By Thomson, manufacturer in Galashiels] Edinburgh, 1831.

This entry is derived from Halkett and Laing's *Dictionary of anonymous and pseudonymous English Literature* new and enl. ed. 1926. We have not seen the book.

mt 14 1832

The hive; a selection of poems, chiefly from the works of living authors. Intended for the perusal of young persons./London, Hailes, 1832./viii, 206 p. front. (of a female young person), 14 cm./A

mt 15 [1840]

The working bee; or Caterer for the hive. Containing a great variety of interesting articles, calculated to blend instruction with amusement./ London, Fisher [1840]/viii, 1535 p. Front. of woman and children with skep, bees and also aeronautical balloon./G

mt 16 1845

The sick bee; a clerico-religio-politico-tragico (a somewhat novel sort of mel-o-drama) poem. By a Brother Bee, impelled thereto from sympathy./ London, Palmer, 1845./v, 83 p./A

An attack on the Church in which the author comments on the words of a sick bee. The idea springs from an entry in Ainsworth's Latin dictionary 'Clerus—the Clergy . . . also a worm, like a spider, which breeds in bee-hives, and makes the bees sick.'

WARREN, Samuel *mt* 17 1851

The lily and the bee./An apologue of the Crystal Palace./London & Edinburgh, Blackwood, 1851./xii, 1 leaf, 224 p./G

Contains a poem, pp. 102–120, addressed to a bee in one of the hives at the Great Exhibition of 1851.

mt 18 1855

The roving bee: or, A peep into many hives. By the author of *Quicksands on foreign shores.* ed. by Mrs Whatley. 2 vol./London, Nisbet, 1855./iv, 463 p. (no p. 222)/A

The history of Dora, a young governess.

mt 19 1856

Honey gathered in commercial rooms by a 'Working Bee'./London, Ward & Lock, 1856./viii, 131 p./A

A paperback addressed 'to commercial travellers' with a picture on the cover of a commercial traveller outside a factory, standing by a skep.

W., H. J. *mt* 20 [1869]

Honey and the honey-comb./London, Macintosh, [1869]./15 p. 12 cm./A

One of H.J.W's Little Books for Children, price one penny. A religious tract.

mt 21 1876

The bees' assistant: a spelling bee guide containing every word in the English language commencing with a silent letter./London, Pitman, 1876./8p./A

BUDGE, Ernest A. Wallis *translator* *mt* 22 1886

The book of the bee. [written by Solomon, Nestorian Bishop at al-Basra c 1222] the Syriac text edited from the manuscripts in London, Oxford, and Munich with an English trans. by Ernest A. Wallis Budge./Oxford, Clarendon Press, 1886. (B)

[LEIGH, Netta *pseud.* **Miss WHITTEMORE]** *mt* 23 [1887]
Busy Bee. by the author of *Dora Selwyn.*/London, Stoneman, [1887]./illus./A

The story of a little girl called Busy Bee.

mt 24 1895
My honey. By the author of *Tip Cat* etc./London, Innes, 1895./354 p./A

mt 25 [1895].
Honey. A religious tract signed W.R.C. [1895]./A

A's copy destroyed by bombing in Second World War.

FRANCE, Anatole *mt* 26 [1911]
Honey-bee./London & New York, John Lane, [1911]./xi, 172 p. 11 pl. front./B

Translated by Mrs. John Lane.

Bee: the princess of the dwarfs./London, J. M. Dent, 1920./159 p. illus. 15 cm.

Translated by Peter Wright.
The Kings Treasuries of Literature.
For children./Hodgson 334.
The story of a princess called Honey-bee.

OXENHAM, John *pseud.*
(DUNKERLEY, William Arthur) *mt* 27 1913
Bees in amber, a little book of thoughtful verse./London, Methuen, 1913./4 leaves, 115 p. 15 cm./

Reprinted many times. Later eds. have 118 p.

mt 28 [1933]
Sweeter than honey. Simple stories illustrating the way of victory, the secret of joy and the worth of true riches./London, Pickering & Inglis, [1933]./A (but copy destroyed by bombing in Second World War.)

(*cancelled*) *mt* 29

SITWELL, Edith *mt* 30 1962
The queens and the hive./London, Macmillan, 1962./

A book about Elizabeth I.

HEYWOOD, Rosalind *mt* 31 1964

The infinite hive: a personal record of extra-sensory experiences. / London, Chatto & Windus, 252 p. 1964. / A

Anr. ed. 1966 London, Pan Books

Anr. ed. 1978 Harmondsworth (Middx.), Penguin Books, 252 p.

The title is taken from Sermon LXX by John Donne (1572–1631): 'This infinite Hive of honey, this insatiable whirlpoole of the covetous mind, no Anatomy, no dissection hath discovered to us . . .'

HERBERT, Frank *mt* 32 1974

Hellstrom's hive. London, New English Library, 1972. / ix, 278 p. Fiction.

Books in special categories

Books about bees written in verse
(author index)

Many of the books with metaphorical titles (*page* 235) *are also in verse*

Books of adult fiction about bees and beekeepers
(author index)

Children's books on bees and beekeeping
(chronological list)

None have been traced before 1800.

Short-title Index

ms = manuscript (before 1500)
mt = metaphorical title (text not about bees)

Hyphens used in titles are retained, but the alphabetical sequence disregards them.

Author Index

ms = manuscript (before 1500)
mt = metaphorical title (text not about bees)
( ) = the person is mentioned in this entry, but is not an author of the book to which the entry relates.

Index of special subjects

The index has been prepared from the titles and notes here, not from the books themselves; many of the subjects below are also dealt with in other more general books. Terms which would have too many entries to be useful (beekeeping, skeps, etc.) are not used in this index.

BRITISH BEE BOOKS

A BIBLIOGRAPHY 1500–1976

International Bee Research Association

Contents

British books that deal with bees and beekeeping are a national, and indeed a world heritage. About 830 printed books are recorded here, and 12 manuscripts written before printed books began, which together cover a span of 1500 years. With a few exceptions they are in English, and constitute the richest collection of material written on the subject in any single language.

The task of finding and recording all these books, annotating the entries about them, and establishing which libraries hold copies, has taken 25 years. It has been a labour of love carried out by bibliophile Members of the International Bee Research Association, first by Joan Harding, a librarian, and Rosamund Duruz who had special historical knowledge, and later by David Smith, a barrister, and Eva Crane, a scientist, who are respectively Secretary and Director of the Association.

The illustrations are reproduced from the books themselves, their dates ranging from A.D. 1000 to 1886.

This book will be invaluable for librarians, booksellers and collectors of bee books. More than this, the many comments and illustrations provide an annotated history of beekeeping which any reader can dip into at will. The indexes include one on special subjects, and this bibliography thus complements *The Bee World. index to Volumes 1–30, 1919–1949* (IBRA 581a), and the *Index to Apicultural Abstracts 1950–1972* (IBRA 604), and provides a valuable insight into the bee research of past generations.

The woodcut reproduced on the cover is the earliest illustration of bees and hives to be printed in England **(The grete herbell, 1526).**

Order form

To: International Bee Research Association, Hill House, Gerrards Cross, Bucks SL9 0NR, England

Please send.................... copy/ies of *British bee books: a bibliography 1500–1976* (£13 or US $30 per copy, post free)

Name ....

Address...

....

....

X62 79.6.1500 UB

D. J. Godfrey. J. D. Cooper

HIVING UNDER DIFFICULTIES.

Fig. 15. Frontispiece of *Modern bee-keeping* published by the B.B.K.A. in 1880 (entry 349). Later illustrations in the book show 'modern' frame hives, but the frontispiece is of interest as a social document. The lady of the house is tanging the bees by hitting a spoon against a dustpan (this was supposed to make the swarm settle). It is the servant girl who climbs up the ladder to dislodge the swarm, and a farmhand who holds a skep to catch the falling bees. No one wears a bee veil.